Eruptions, Initiatives and Evolution in Citizen Activism

Eruptions, Initiatives and Evolution in Citizen Activism is the result of a collaborative research project spanning Asia, Africa, Latin America, and Europe. The book analyses internal and external challenges to civil society in more than twenty countries. It investigates through studies of countries that include South Africa, India, and the Netherlands civil society evolution; examinations of citizen activism, such as Occupy London, the Chilean student movement, the Cambodian farmers campaign against land grabs; regional overviews such as the Southern Cone of Latin America, Southern Africa, and Russia. The studies identify changing roles, capacities, contributions and limitations of civil society in response to changing political, economic, and social contexts. The book goes on to present selected studies, identifies patterns and lessons that emerge across countries and regions. It articulates the implications of those lessons for practitioners and policy makers concerned with civil society contributions to national and regional development.

This book was originally published as a special double issue of *Development in Practice*.

Rajesh Tandon is the founder President of the Society for Participatory Research in Asia (PRIA), an India-based NGO that provides civil society research and capacity building. He has been active in regional and international initiatives to strengthen civil society, democratic governance, and active citizenship.

L. David Brown is a retired Senior Research Fellow at the Hauser Center for Nonprofit Organizations at Harvard University. He was co-Founder and President of the Institute for Development Research, a civil society research and capacity building center in Boston. He works on civil society development, cross-sector collaboration, and transnational civil society organisations.

Development in Practice Books
Series Editor: Brian Pratt

Each title in the *Development in Practice Books* series offers a focused overview of practice-relevant analysis, experience, and research on key topics in development.

Participatory Research and Gender Analysis
New Approaches
Edited by Nina Lilja, John Dixon, and Deborah Eade

Achieving Education for All through public–private partnerships?
Non-state provision of education in developing countries
Edited by Pauline Rose

Global Food Price Shocks and Poor People
Themes and Case Studies
Edited by Marc J. Cohen and Melinda Smale

Child Protection in Development
Edited by Michael Bourdillon and William Myers

Religion, Religious Organisations and Development
Scrutinising religious perceptions and organisations
Edited by Carole Rakodi

Eruptions, Initiatives and Evolution in Citizen Activism
Civil Societies at Crossroads
Edited by Rajesh Tandon and L. David Brown

Eruptions, Initiatives and Evolution in Citizen Activism

Civil Societies at Crossroads

Edited by

Rajesh Tandon and L. David Brown

Routledge
Taylor & Francis Group
LONDON AND NEW YORK

INTRAC
International NGO Training and Research Centre

First published 2015 by Routledge

2 Park Square, Milton Park, Abingdon, Oxon, OX14 4RN
605 Third Avenue, New York, NY 10017

Routledge is an imprint of the Taylor & Francis Group, an informa business

First issued in paperback 2020

British Library Cataloguing in Publication Data
A catalogue record for this book is available from the British Library

ISBN 13: 978-1-138-80481-4 (hbk)
ISBN 13: 978-0-367-73966-9 (pbk)

Typeset in Times New Roman
by RefineCatch Limited, Bungay, Suffolk

Publisher's Note
The publisher accepts responsibility for any inconsistencies that may have arisen during the conversion of this book from journal articles to book chapters, namely the possible inclusion of journal terminology.

Contents

Citation Information

The chapters in this book were originally published in *Development in Practice*, volume 23, issue 5–6 (August 2013). When citing this material, please use the original page numbering for each article, as follows:

Chapter 1
Guest Editors' Introduction: Civil societies at crossroads: eruptions, initiatives, and evolution in citizen activism
Rajesh Tandon and L. David Brown
Development in Practice, volume 23, issue 5–6 (August 2013) pp. 601–608

Chapter 2
Struggles for systems that nourish: southern Africa civil society contributions and challenges to the creation of flourishing societies
Mutizwa Mukute and James Taylor
Development in Practice, volume 23, issue 5–6 (August 2013) pp. 609–616

Chapter 3
Struggles against systems that impoverish: South African civil society at the crossroads
James Taylor
Development in Practice, volume 23, issue 5–6 (August 2013) pp. 617–630

Chapter 4
Kampala city traders (KACITA) strike for action
Jackline Kabahinda
Development in Practice, volume 23, issue 5–6 (August 2013) pp. 631–643

Chapter 5
Civil society at multiple crossroads in Asia
Kaustuv Kanti Bandyopadhyay
Development in Practice, volume 23, issue 5–6 (August 2013) pp. 644–652

Chapter 6
Civil society in changing India: emerging roles, relationships, and strategies
Debika Goswami and Rajesh Tandon
Development in Practice, volume 23, issue 5–6 (August 2013) pp. 653–664

Chapter 7
Changing civil society in Cambodia: in search of relevance
Kaustuv Kanti Bandyopadhyay and Thida C. Khus
Development in Practice, volume 23, issue 5–6 (August 2013) pp. 665–677

Chapter 8
Is civil society in the Southern Cone of Latin America at a crossroad?
Anabel Cruz
Development in Practice, volume 23, issue 5–6 (August 2013) pp. 678–684

Chapter 9
Protest and proposal, participation and representation: the Chilean student movement, 2011–12
Inés M. Pousadela
Development in Practice, volume 23, issue 5–6 (August 2013) pp. 685–700

Chapter 10
From embarrassing objects to subjects of rights: the Argentine LGBT movement and the Equal Marriage and Gender Identity laws
Inés M. Pousadela
Development in Practice, volume 23, issue 5–6 (August 2013) pp. 701–720

Chapter 11
The emergence and re-emergence of civil society: a brief history of civil society in Europe, from the Magna Carta to the Eurozone crisis
Brian Pratt and Rowan Popplewell
Development in Practice, volume 23, issue 5–6 (August 2013) pp. 721–728

Chapter 12
Occupy London as pre-figurative political action
Neil Howard and Keira Pratt-Boyden
Development in Practice, volume 23, issue 5–6 (August 2013) pp. 729–741

Chapter 13
Dutch civil society at crossroads
Rik Habraken, Lucas Meijs, Lau Schulpen, and Cristien Temmink
Development in Practice, volume 23, issue 5–6 (August 2013) pp. 742–754

Chapter 14
Treading new ground: a changing moment for citizen action in Greece
Maro Pantazidou
Development in Practice, volume 23, issue 5–6 (August 2013) pp. 755–770

Chapter 15
Russian civil society: background, current, and future prospects
Charles Buxton and Evgenia Konovalova
Development in Practice, volume 23, issue 5–6 (August 2013) pp. 771–783

Chapter 16

Civil societies at crossroads: lessons and implications
Rajesh Tandon and L. David Brown
Development in Practice, volume 23, issue 5–6 (August 2013) pp. 784–796

Please direct any queries you may have about the citations to
clsuk.permissions@cengage.com

INTRODUCTION

Civil societies at crossroads: eruptions, initiatives, and evolution in citizen activism

Rajesh Tandon and L. David Brown

This Special Issue has grown from the sense that important changes in the last two decades pose dilemmas and challenges for civil societies in many countries. The Issue reports on a series of studies of the evolving roles of civil society sectors and citizen initiatives in several regions of the world. This introduction identifies a series of events and forces that over the last two decades have fundamentally changed the contexts of civil society activities in many countries. These changes have catalyzed a wide range of citizen eruptions and initiatives on particular issues as well as national civil society evolutions in many countries. The papers in this Special Issue have resulted from a multi-country collective reflection organized by five civil society support organizations from different regions. They sought to identify roles, capacities, contributions, and limitations of civil society in these changing contexts using a variety of approaches to data collection and analysis. This introduction briefly describes the papers in the Special Issue. They include regional overviews, descriptions of national sector evolution, and cases of citizen activism in Southern and Eastern Africa, Asia, Southern Latin America, Western Europe and Russia. The final paper provides an overview of the lessons learned from comparative analysis across these and other cases and draws some of the implication of those lessons for practitioners and policy makers.

La société civile à la croisée des chemins : éruptions, initiatives et évolution de l'activisme citoyen
Ce Numéro spécial est né du sentiment que les changements importants qui se sont produits au cours des deux dernières décennies soulèvent des dilemmes et des défis pour les sociétés civiles de nombreux pays. Ce numéro présente une série d'études sur les rôles en évolution des secteurs de la société civile et des initiatives citoyennes dans plusieurs régions du monde. Cette introduction identifie une série d'événements et de forces qui, au cours des deux dernières décennies, ont fondamentalement modifié les contextes des activités de la société civile dans de nombreux pays. Ces changements ont catalysé un large éventail d'éruptions et d'initiatives citoyennes sur ces questions particulières, ainsi que des évolutions de la société civile dans de nombreux pays. Les articles figurant dans ce Numéro spécial ont découlé d'une réflexion collective multi-pays organisée par cinq organisations de soutien à la société civile de différentes régions. Elles ont cherché à identifier les rôles, les capacités, les contributions et les limites de la société civile dans ces contextes en mutation à l'aide d'une variété d'approches de la collecte et de l'analyse des données. Cette introduction donne une brève description des articles du Numéro spécial. Ils regroupent des aperçus régionaux, des descriptions de l'évolution du secteur national et des cas d'activisme citoyen en Afrique australe et de l'Est, en Asie, dans le sud de l'Amérique latine, en Europe de l'Ouest et en Russie. Le dernier article donne

un aperçu des enseignements tirés d'une analyse comparative dans tous ces cas, et d'autres, et soulève certaines des implications de ces enseignements pour les praticiens et les décideurs.

Sociedades civiles en la encrucijada: erupciones, iniciativas y evolución del activismo ciudadano
El presente Número Especial se originó a partir de la idea de que los importantes cambios ocurridos durante las últimas dos décadas presentan dilemas y retos para las sociedades civiles de muchos países. Este número da cuenta de una serie de estudios realizados en torno a los roles en evolución de los sectores de la sociedad civil y a sus iniciativas en diversas regiones del mundo. La presente introducción identifica una serie de eventos y de fuerzas que, durante los últimos veinte años, ha cambiado sustancialmente los contextos en los cuales se llevan a cabo las actividades de la sociedad civil en muchos países. Estos cambios han impulsado una amplia gama de erupciones y de iniciativas ciudadanas en relación con temas específicos, así como de evoluciones de la sociedad civil en un buen número de países. Los artículos reunidos en este Número Especial surgen de las reflexiones colectivas realizadas en diversos países por cinco organizaciones de apoyo de la sociedad civil de distintas regiones. Apoyándose en varios enfoques para la recopilación de información y la elaboración de análisis, dichas reflexiones pretendieron identificar roles, capacidades, contribuciones y limitaciones de la sociedad civil en estos contextos cambiantes. La presente introducción resume los artículos de este Número Especial. Los mismos incluyen descripciones regionales, descripciones de la evolución de los sectores nacionales y casos de activismo ciudadano en África del Sur y del Este, en Asia, en Sudamérica, en Europa Occidental y en Rusia. El último artículo ofrece una descripción general de los aprendizajes surgidos de los análisis comparativos obtenidos a partir de estos y de otros casos, examinando algunas de las implicaciones que dichos aprendizajes han tenido para los operadores y los formuladores de políticas.

Sociedades civis na encruzilhada: eclosões, iniciativas e evolução no ativismo dos cidadãos
Esta edição especial foi desenvolvida a partir do senso de que mudanças importantes nas últimas duas décadas têm apresentado dilemas e desafios para as sociedades civis em vários países. A Edição relata uma série de estudos dos papeis cada vez mais importantes dos setores da sociedade civil e das iniciativas dos cidadãos em várias regiões do mundo. Essa introdução identifica uma série de eventos e forças que durante as duas últimas décadas têm fundamentalmente mudado os contextos das atividades da sociedade civil em vários países. Essas mudanças têm catalisado uma ampla série de eclosões e iniciativas de cidadãos sobre questões particulares e também, nacionalmente, evoluções da sociedade civil em vários países. Os artigos desta Edição Especial são resultado de uma reflexão coletiva de vários países organizada por cinco organizações de apoio à sociedade civil de diferentes regiões. Eles buscaram identificar os papéis, capacidade, contribuições e limitações da sociedade civil nestes contextos variáveis utilizando várias abordagens para coleta e análise de dados. Essa introdução descreve brevemente os artigos da Edição Especial. Eles incluem visões panorâmicas regionais, descrições da evolução do setor nacional e casos de ativismo de cidadãos no sul e leste da África, Ásia, sul da América Latina, Europa Ocidental e Rússia. O artigo final oferece uma visão geral das lições aprendidas a partir de uma análise comparativa entre esses e outros casos e extrai algumas das implicações destas lições para profissionais e formuladores de políticas.

Introduction

Civil societies at crossroads? This Special Issue has grown from the sense that important changes in the last two decades are posing central dilemmas and challenges for civil societies in many countries. The Issue reports on a series of studies of the evolving roles of civil society sectors and citizen initiatives in several regions of the world, and seeks to identify major themes visible in these stories and their implications for practitioners and policymakers.

Two decades ago a series of events catalysed changes that amounted to a new world order. The fall of the Berlin Wall in 1989 reflected major geopolitical shifts and the end of the Cold War. The next decade saw the cresting of the "Third Wave" of democratisation (Huntington 1992), initially in Latin America and later in Eastern Europe and other regions. That wave catalysed widely-visible challenges to many autocratic regimes, such as the Tiananmen Square demonstrations. Globalisation and economic liberalisation in countries like China and India has brought millions out of poverty, but it has also emphasised the growing gap between the rich and the poor. At the same time, civil society emerged in many different countries as a major actor in social, political, and economic development, prompting one observer to suggest that the "*associational revolution*" was as important as the earlier rise of the nation state (Salamon 1994).

Civil society as we use the term includes a wide range of social organisations that mobilise resources around values and voluntary social energy. They include non-profit and non-governmental organisations, unions, churches, neighbourhood associations, soccer clubs, social movements, and a wide variety of citizen initiatives. One key element of civil societies in many countries has been non-governmental development organisations (NGOs) that are formally organised to work with poor and marginalised groups. Some NGOs are domestic; other are international (INGOs). In recent years many civil society initiatives have taken less formally organised forms such as networks, informal alliances, and social movements.

Today new forces appear to be challenging the perspectives and capacities of civil societies that evolved to respond to that earlier world order. While many democracies have persisted, the economic successes of non-democratic regimes in China and Singapore have raised questions about the link between democracy and economic growth. Economic power appears to be shifting from North America and Europe to China and India. The rise of the private sector and neo-liberalism has raised questions about the viability of the welfare state model, particularly since the financial crisis of 2008. In many countries democratic politics have been undermined or captured by concentrations of wealth and power in a few global elites. The revolution in information and communications technology has vastly increased the capacity of ordinary citizens to get information about events in their own and others' countries and to coordinate initiatives with each other. All these combine with expanding economies that demand more of increasingly-scarce resources, such as water, arable land, and stable climates, that are needed to feed growing populations with rising expectations.

In this context citizens in many countries are questioning distributions of power and wealth and its abuse – witness the Arab Spring in many countries, the Occupy and "indignado" movements in Europe and North America, student challenges to education in Chile and Canada, citizen uprisings against corrupt politics in India, Myanmar, and Malawi. Citizens increasingly distrust social and political institutions and challenge the unaccountable use of power around the world.

While civil society actors remain more trusted than many other institutions, public faith in their commitment and capacity to promote the public good apparently rests on fragile foundations (Globescan 2012). While some remain optimistic about the roles of civil society activism in general (Hawken 2007), others suggest that many NGOs and formal civil society organisations/actors remain unconnected and marginally legitimate with the less formal citizen initiatives that have emerged in many contexts (CIVICUS 2011).

In 2012, a number of civil society support organisations came together in to work on a collective reflection process about the future of civil society around the world. The partners in the "Civil Society at Crossroads?" initiative – PRIA in India, PSO in the Netherlands, INTRAC in the UK, CDRA in South Africa, EASUN in Tanzania, and ICD in Uruguay – agreed to examine the challenges facing civil society and the responses emerging in their regions. They agreed to explore two kinds of patterns: (1) examples of national civil society *sector evolution* as their contexts evolve, and (2) examples of *citizen activism*, including eruptions of citizen activity that pose unexpected challenges to existing social orders, and citizen initiatives to solve problems that use existing systems to pursue new priorities.

The central focus of the inquiry has been to identify roles, capacities, contributions, and limitations of civil society in these changing local and global contexts. More specifically the Crossroads partners have tried to identify emerging local formations, their relations with other civil society actors, their impacts on poor and excluded groups, and their relations with the state and market sectors. They looked at changes in economic resources and political space for civil societies, and at similarities and differences across poor, middle-income, and affluent countries. They also sought to explore the emergence of hybrid formations and the impacts of change across national, regional, and transnational boundaries.

Data collection and analysis

This initiative explores these emerging dynamics to develop concepts and hypotheses rather than to test hypotheses from an already developed theory. We have focused on trying to understand stories across countries and regions through comparative analysis (Miles and Huberman 1994; Yin 1984). We have used a range of data collection methods to explore our questions; some focused on evolutions of civil society as a sector, others aimed at assessing regional patterns of development, and still others emphasised citizen eruptions or initiatives. We have convened forums with civil society practitioners, developed case studies of sector evolution and citizen activism, organised discussions of cases and findings with civil society experts, and used comparative analysis to look at patterns across cases.

In this Special Issue we present three kinds of analysis: regional viewpoints across a number of countries that share some socio-economic and cultural contexts, analyses of national civil society sector development, and stories of citizen activism that have emerged as civil societies have reacted to changing contexts. Table 1 summarises these contributions.

Table 1: Sector evolution, citizen activism, and regional viewpoints

Region	*Sector development*	*Citizen activism*	*Regional viewpoints*
Africa	• South Africa	• Uganda city traders' strike	• Southern Africa
Asia	• India	• Cambodia land grab	• Asia
Europe	• The Netherlands	• Greek Syntagma Square • UK Occupy London	• Western Europe
Latin America		• Argentina equal marriage • Chile student movement	• Southern Cone of Latin America
Other	• Russia		

We have selected materials for the Special Issue from more than 20 cases generated by the Crossroads project. We chose cases that reflect sector evolution and citizen activism and that represent diversity across regions.

The papers

The papers in the Special Issue are organised by regions so that readers can examine in one place regional viewpoints, national sector evolution, and citizen activism that have emerged from similar contexts. In this section, we briefly summarise those contributions.

Southern and Eastern Africa

James Taylor and Mutizwa Mukute provide a viewpoint of the evolutions of civil societies in the Southern African region. Civil society actors have been prominent in many countries in the region, though they often have difficulty in influencing the activities of other powerful sectors. Struggles with the system in South Africa, efforts to hold Presidents who abuse their power accountable in Malawi, and campaigns to limit violence and abuses of power in Zimbabwe have tested and often exceeded the capacities of civil society actors. Many countries in Southern Africa continue to struggle to meet the expectations of their peoples and to build and maintain institutions that enable effective democratic governance in the face of concentrated power.

The evolution of civil society in South Africa is described by **James Taylor**, who has been an activist in that evolution for 30 years. He traces civil society resistance to the apartheid system and its participation in the effort to build a more equitable multi-racial society over the last two decades. He suggests that civil society may have come full circle, from advocacy against the regime, to participation in building a new society, to disillusionment with the difficulty in changing "systems that impoverish" most of their members. He calls for new thinking at the crossroads facing South African civil society that might connect with a "developmental state" and a "social business" to construct a society better able to meet the needs of all its members.

Jackline Kabehinda describes a strike by the Traders Association in Uganda to press the government to change Ministry of Finance decisions to retroactively raise interest rates. The Association organised a strike that expanded from one to several days, and gained support from a wide variety of other stakeholders. While the strike succeeded in bringing the government to the table to talk with Association representatives, it is less clear that it succeeded in compelling more than cosmetic changes in government policy. But the strike also mobilised growing commitment from middle-class constituents to challenge policy and advocate with the government, and so reflects emerging activism among Ugandan citizens other than poor and marginalised groups.

Asia

Kaustuv Bandhopadhyay offers a viewpoint on the evolution of civil societies in Asia, grounded in a series of conferences and debates in several countries. The paper describes "Asia on the rise" in economic terms, which has led to dramatic increases in inequality and corruption in many countries. Civil societies have made important contributions to serving the poor, innovations in participatory development, empowerment of marginalised groups, fostering actives citizenship, enabling demand for reform, and policy discourses about public goods. But the changing context has created a number of crossroads for civil societies, including the challenges of recognition or repression, relations with governments, relations with political societies, diminishing financial support and political space, and expanding international challenges. In the future, civil societies in the region must grapple with protecting their political space, building wider coalitions, mobilising new resources, and engaging with new international actors.

In the last 30 years there has been a dramatic shift to liberalisation, privatisation, and globalisation in India, and **Debika Goswami and Rajesh Tandon** describe the changes associated with that shift and their implications for civil society roles, relationships, and strategies. Although civil society has a long history in India, the reforms of the early 1990s catalysed economic growth, governance decentralisation, spreading information and communication technology, and variety of other factors that affected civil society actors. Civil society roles became more diversified; relations with state and market institutions evolved to include both cooperation and confrontation; social movements emerged to challenge long-standing problems like corruption; resources from international sources began to shrink. At the same time, the need for civil society as a counterweight to other powerful institutions has never been greater.

The evolution of civil society in the Cambodian transition from the Khmer Rouge era is the focus of the analysis of **Kaustuv Bandhopadhyay and Thida C. Khus.** CSOs have been important actors in democracy-building and economic development, heavily influenced by resources and ideas from international development aid. International NGOs have fostered the emergence of local CSOs to deliver services; trade unions are emerging in the modernised sectors of economy; peacebuilding CSOs are giving way to service delivery and livelihood CSOs; human rights and government accountability CSOs remain controversial in the eyes of government agencies. Relations with the government draw most attention now, but linkages among CSOs, movements, and business organisations are expected to become increasingly important, as will alliances with media and academia and initiatives to hold government and business accountable for their impacts on grassroots populations.

Southern Latin America

Anabel Cruz provides an overview of civil societies in the southern Latin American region. The region underwent a political transformation in the last 20 years, moving from authoritarian regimes to multi-party democracies. NGOs and other civil society organisations played an important part in that transition, but now new forms of citizen activism are emerging. Some are social movements that challenge existing regimes and institutions, as in the Chilean student movement; some are political campaigns for new rights and legislation, like the campaigns for equal marriage in Argentina and for legal abortions in Uruguay; still others are networks of activists, like the regional youth movement to build housing for the poor. With the decline of foreign funding and the rise of democracies, emerging civil society initiatives must be more grounded in the concerns of local peoples than the priorities of international donors, and they must become influential in national and regional as well as local contexts.

In her analysis of the Chilean student movement **Inés M. Pousadela** emphasises its evolution into a campaign aimed at an inequitable system inherited from the Pinochet dictatorship. The student movement used a wide range of creative strategies to confront the government, gradually building a national alliance that included faculty, parents, unions, and many other stakeholders in the national educational system. Although there is a substantial history of citizen activism in Chile, this campaign was one of the first to emerge independently of all the political parties and it mobilised the largest demonstrations since the overthrow of the dictatorship more than 20 years ago.

The campaigns to pass the Equal Marriage and Gender Identity Laws in Argentina were so successful that they astonished even the citizen leaders who planned and implemented them. **Inés M. Pousadela** reports on the evolution of the campaign and its rapid reshaping of public attitudes as well as building legislator support for the new laws in spite of strong resistance from the hierarchy of the Catholic Church, long dominant in areas of religious teachings. The leadership of the Lesbian Gay Bisexual and Transsexual Alliance built on similar legislation passed in Spain and evolving public attitudes that supported marriage by couples of the same sex as well as personal decisions about gender identity. The campaign dramatically changed the options available to Argentine citizens.

Western Europe

In their European regional viewpoint **Brian Pratt and Rowan Popplewell** explore the deep historical roots of European civil societies and the more recent growth of state support for their services. That support has created the current crises of financial resources and popular relevance for civil societies that have become too dependent on state funding. The global financial crisis and changing roles of the state have reduced funding for civil society and catalysed citizen activism that has been largely unconnected to existing civil society organizations. They note the growing importance of under the radar voluntary organisations and the emergence of so-called uncivil society as themes that may become increasingly important in the future.

The emergence of Occupy London as so-called prefigurative politics is assessed by **Neil Howard and Keira Pratt-Boyden,** who were both participant observers in the movement. They discuss Occupy's genesis in the financial crises and the widespread criticism of its failure to articulate clear strategies and concrete demands. They argue that Occupy sought rather to create a space in which participants could be the change they want and engage in the kinds of democratic participation that are rare or even impossible in capitalist societies. They describe in some detail how those concerns were expressed in three aspects of Occupy: the General Assembly that undertook governance issues; the Tent City University, in which "*anyone can teach and everyone can learn*"; and the Welfare and Well-being Group that provided support to all comers, many of whom were drawn to the movement by the prospect of free food and human support.

In describing the evolution of Dutch civil society**, Rik Habraken, Lucas Meijs, Lau Schulpen, and Cristien Temmink** describe five historical phases of segmentation of the non-profit sector according to religion and class differences supported by government resources, giving way to liberalisation and privatisation, and then more recently the withdrawal of state subsidies. They describe the impacts of this evolution for a deprived neighbourhood community and for a national non-profit refugee council. They argue that the Dutch experience suggests important lessons for how civil society relations with governments can influence popular support for civil society institutions, their capacity for advocacy, and government perceptions of civil society.

Maro Pantazidou examines the eruption of citizen activism in Greece that emerged in response to austerity measures and the violations of democratic process that characterised government response to the debt crisis. Both the December riots in 2008 and the Syntagma Square Movement in 2009 expressed citizen anger at the implosion of the Greek economy and the erosion of democracy. Those initiatives have generated citizen commitment and experiments in solidarity, direct democracy, unruly actions in response to unfair institutions, and efforts to create more autonomous self-governance. The initiatives launched in mass public demonstrations have now begun to permeate a wide variety of other contexts in which citizens seek participation and influence, reflecting the potential emergence of new forms of citizen action and social change.

Other regions

While it has not been possible to cover all the regions in which civil societies are at crossroads, we have one other regional viewpoint represented here. **Charles Buxton and Evgenia Konovalova** describe the evolution of civil society in Russia. They argue that although some analysts regard civil society as a new phenomenon in Russia, others have identified civil society development that has evolved in several stages since the mid-1700s. They provide brief historical summaries and typical case studies to illustrate four earlier stages. They offer a more detailed discussion of Russian civil society today, supplemented by case descriptions for regions and for the anti-globalisation movement. They suggest that civil society is more active in Russia than many outsiders recognise, increasingly engaged in contracting for service delivery with local governments and in consultations with other sectors around issues of local and national concern.

And now...

This introduction sets the context for the Civil Societies at Crossroads stories that follow. We turn now to the regional viewpoints and the stories of sector evolution and citizen activism from those regions.

In the conclusion we will explore themes and lessons that emerge from comparison of these and other stories developed by the Crossroads initiative. We will identify what we see as the most important lessons to be learnt across the stories and some implications of those lessons for civil society practitioners and for policymakers concerned about civil society contributions to sustainable development in the future.

References

CIVICUS. 2011. *The State of Civil Society 2011*. Johannesburg: CIVICUS World Alliance for Citizen Participation.

Globescan. 2012. "High Public Trust in NGOs, but Is It Built on Shaky Foundations?" *Globescan*. Accessed October 14. www.globescan.com/commentary-and-analysis/featured-findings/entry/high-public-trust-in-ngos-but-is-it-built-on-shaky-foundations.html

Hawken, P. 2007. *Blessed Unrest: How the Largest Movement in the World Came into Being and Why No One Saw It Coming*. New York: Viking.

Huntington, S. P. 1992. *The Third Wave: Democratization in the Late Twentieth Century*. Norman: University of Oklahoma Press.

Miles, M. B., and A. M. Huberman. 1994. *Qualitative Data Analysis*. 2nd ed. Thousand Oaks, CA: Sage.

Salamon, L. M. 1994. "The Global Associational Revolution: The Rise of the Third Sector on the World Scene." *Foreign Affairs* 73: 111–124.

Yin, R. K. 1984. *Case Study Research*. Beverly Hills, CA: Sage.

The authors

Rajesh Tandon (corresponding author) is founder-President of PRIA, New Delhi. Over the past three decades, under his leadership, PRIA has extensively engaged in building the capacities of civil society organisations around the world. Rajesh has an electronics degree from IIT (Kanpur), an MBA from IIM (Calcutta), and a PhD in Organisational Behaviour from CWRU (Cleveland). He provided foundational leadership to International Forum of Capacity Building of Southern NGOs (1997–2001), CIVICUS (1993–1999), FIM-Forum for Democratic Global Governance (1998–2011), and Voluntary Action Network India (1988–2008). He has written extensively, with manuals, research papers, books, and training materials on civil society, capacity development, participatory research, and democratic governance.

L. David Brown is semi-retired as Senior Research Fellow at the Hauser Center for Nonprofit Organizations at Harvard University. He has been President of the Institute for Development Research, a nonprofit centre for development research and consultation, and taught at schools of management at Boston University and Case Western Reserve University as well as the Harvard Kennedy School of Government. His research and consulting focuses on civil society organisations, inter-organisational networks, and multi-sector partnerships that foster sustainable development, accountable governance, and social justice around the world.

Struggles for systems that nourish: southern Africa civil society contributions and challenges to the creation of flourishing societies

Mutizwa Mukute and James Taylor

This paper discusses the concept, contexts, evolution, contributions, and challenges of civil society in the southern Africa based on three case studies of Malawi, South Africa, and Zimbabwe. It concludes by identifying common patterns and their implications for civil society and civil society organisations in southern Africa.

Luttes pour des systèmes qui nourrissent : contributions de la société civile en Afrique australe et défis pour la création de sociétés prospères
Ce document traite du concept, des contextes, de l'évolution, des contributions et des défis de la société civile en Afrique australe sur la base de trois études de cas menées au Malawi, en Afrique du Sud et au Zimbabwe. Il se conclut en identifiant des schémas communs et leurs implications pour la société civile et les organisations de la société civile en Afrique australe.

Luchas a favor de sistemas de fomento: contribuciones y desafíos enfrentados por la sociedad civil de África del Sur para crear sociedades prósperas
Centrándose en tres estudios de caso realizados en Malawi, Sudáfrica y Zimbabue, el presente artículo examina el concepto, los contextos, la evolución, las contribuciones y los retos que enfrenta la sociedad civil en el sur de África. Finaliza identificando los patrones en común y las implicaciones que éstos tienen tanto para la sociedad civil como para sus organizaciones en el sur de África.

Lutas por sistemas que alimentam: contribuições da sociedade civil do sul da África e desafios para a criação de sociedades prósperas
Este artigo discute o conceito, contextos, evolução, contribuições e desafios da sociedade civil no sul da África baseado em três estudos de caso de Malauí, África do Sul e Zimbábue. Ele conclui identificando modelos comuns e suas implicações para a sociedade civil e organizações da sociedade civil no sul da África.

Introduction

This paper explores the role of the social and economic justice aspect of civil society in Southern Africa at a time when global economic and political forces are re-aligning around the need to access resources to fuel economic growth. It links the development of this aspect of civil society to the history of countries in the region: a history shaped by previous extractive regimes that have concentrated (and exported) wealth while impoverishing the majority of indigenous populations. It posits that the history of the region is important at this time as a reminder of who pays the real cost of economic development. The paper looks at how civil society responds to this injustice and to the impoverishment of human and ecological systems that accompany it. It ends by raising questions around the challenges facing civil society in its ongoing attempts to establish systems that nourish as opposed to impoverish.

Although not an exhaustive study of the whole region, in this paper southern Africa refers to the region that includes the 15 member states of the Southern African Development Community (SADC). While the major focus is on civil society activities over the last 20 years, the paper also covers prior developments that have given rise to the contemporary issues that civil society is responding to. It highlights the civil society organisations' (CSOs) responses consolidating around human rights and good governance; and economic justice associated with the redistribution of productive resources. It links the current structural economic injustice to colonisation and the resultant dispossession and displacement of indigenous peoples. This continued after independence as the former master said to the black man, "*You take the crown and I will keep the jewels*" (Mandaza 1999, 81). The established systems of wealth extraction and concentration have been sustained partly because of present day corruption by the elite (Pellicer et al. 2011) and a new wave of land grabbing (Palmer 2008).

Human rights and governance issues have arisen from the entrenched culture of violence and impunity that was cultivated during colonial conquest, and continued through the military liberation struggles, and authoritarianism and (self-righteous) intolerance by post-independence governments (Edwards 2009; Sachikonye 2012). Strong traditional hierarchical and conservative systems in countries like Swaziland further result in human rights abuses that civil society is driven to confront.

The concept of civil society

Generally civil society refers to the arena outside the family, the state, and the market which is created by individual, voluntary, and collective actions to advance shared interests (Belay, Firmin, and Pegus 2012). This paper focuses on the formal and informal, traditional and modern, secular and religious actions of civil society, aimed at the development of a society with greater social and economic equity and democracy (Edwards 2009). Civil society works towards achieving this by promoting accountability, democracy, and the rights of citizens, and creating alternative solutions to the distribution of power and resources (Beauclerk, Pratt, and Judge 2011). It exists alongside and in relation to other societal systems such as government and business (Perlas 2000).

In southern Africa CSOs are a heterogeneous community that includes humanitarian charities, trusts, non-governmental organisations (NGOs), community based organisations (CBOs), women's organisations, faith-based organisations (FBOs), professional associations, trade unions, credit and burial associations, self-help groups, social movements, neighbourhood associations, business and economic empowerment associations, youth organisations, student unions, farmer organisations, coalitions, independent media, independent research institutes, and advocacy groups (Kaapama et al. 2007; Edwards 2009). The concept excludes political parties.

Key societal issues and civil society responses

The nature of civil society and its interventions is determined by context. This paper identifies three interrelated phases that created different contexts for civil society in southern Africa: colonial; decolonisation; and independence. Civil society contributions in these contexts inform our conclusion that it has struggled for systems that nourish to replace those that impoverish. It is worth noting that the evolution and associated contributions of civil society are not neatly bound, but rather iterative and messy.

Colonial period and the struggle for racial justice

During colonial and minority rule, governments and the business sector worked together to promote systems that enriched the minority and impoverished the local populations under different forms of racial discrimination. Civil society in the region responded by tackling racial discrimination and associated policies and laws. For example, in Zimbabwe, in the 1950s the Inter-Racial Association of Southern Rhodesia was formed to address inter-racial issues while the workers unions such as the African Workers' Voice Association embarked on strikes and demonstrations for improved wages and working conditions (Mlambo 2009). The net effect of the colonial period on present-day issues that civil society has to deal with is the skewed distribution of productive resources, which are largely in the hands of non-indigenous peoples (Shivji 2007).

Decolonisation and the struggle for majority rule

During decolonisation, which took place from the 1960s to 1994, southern African countries used three avenues to gain independence and majority rule: peaceful negotiation and handover involving little or no violence; armed struggle; and mass resistance. The responses of civil society in these different circumstances varied, but the unmet rights and needs of the oppressed were a significant driver in the organising of civil society. One of the current negative issues that emerged from the decolonisation process is a culture of violence and impunity that accompanied liberation struggles (Bourne 2011), which has given rise to the need for conflict transformation and reconciliation (Colvin 2007). African nationalism and armed struggle bred intolerance of opposing voices and nurtured authoritarianism (Sachikonye 2012). During the negotiations and struggles for independence, civil society's contribution included the provision of humanitarian assistance, civic education, and building political consciousness. CSOs that played a key role during this era included women's clubs such as *Jekesa Pfugwa* (enlighten the mind), in Zimbabwe; FBOs such as Catholic Commission of Peace and Justice (CCJP); and trade unions such as the Congress of South African Trade Unions (COSATU), in South Africa. Decolonisation processes had the negative effect of perpetuating cultures of violence, impunity, and intolerance that civil society must deal with today.

Independence, transitional justice, expansion of political freedoms and economic justice

In the last 20 years, nearly all SADC member states have achieved independence: countries such as Madagascar, Zambia, Malawi, Botswana, and Swaziland gained independence in the 1960s; Angola and Mozambique in the 1970s; and Zimbabwe, Namibia, and South Africa in 1980, 1990, and 1994 respectively. For most southern African countries, the immediate post-independence period was a honeymoon period because people believed that the new governments would deliver on the people's aspirations. Consequently, CSOs tended to collaborate with majority rule governments. They continued to provide humanitarian assistance, capacity development, and engaged in developmental work to support government initiatives. In some countries they went further and provided input in peace and reconciliation initiatives.

During the post-independence period disillusionment emerged with majority governments that did not deliver on the liberation promises on economic and political emancipation. For example, in Zimbabwe at independence, 67 per cent of national income and the bulk of the resources belonged to 3 per cent of the population, mostly white farmers and a small black bourgeoisie, and in the nine years after independence, 97 per cent of bank loans went to white-owned businesses (Orner and Holmes 2011, 479). In South Africa economic power remains in the hands of a small (largely unchanged) elite with the gap between rich and poor one of the highest in the world and growing (Pellicer et al. 2011). In general these governments made improvements in social services such as education, health, and housing; and ending systems of racial discrimination. Governments' limited achievements have been linked to the uneven international playing field, difficulties in managing structural adjustment programmes, corruption by the political elite and different forms of favouritism, and inability to redistribute national wealth fairly (Pellicer et al. 2011; Sachikonye 2012). Some governments such as those of Malawi, Swaziland, Democratic Republic of Congo (DRC), and Zimbabwe have had challenges in good governance and human rights. People's lack of access to land and mineral resources is a commonly recurring issue in the region. Consequently a vital part of civil society action is currently concerned with constitutional, political, electoral, and economic reforms, including in some countries reforms of the judiciary and security, and reconciliation and conflict prevention (Godsäter and Söderbuam 2008; Church and Civil Society Forum 2010).

Notable contributions of southern African civil society

Focusing on civil society contributions over the last 20 years, during which all southern African states were independent, these contributions can be organised around (a) expanding political freedoms; and (b) increasing access to productive resources.

Struggles to expand political freedoms

One of the most important areas in which CSOs have made a meaningful contribution is in constitutional reform and protecting gains made in that area. For example, in South Africa CSOs participated in drafting the constitutional framework for the new government. In Zimbabwe the National Constitutional Assembly (NCA), and the Church and Civil Society Forum have contributed to the rewriting of the constitution (Church and Civil Society Forum 2010). Civil society in Malawi also successfully used mass protests to stop the second post-colonial President from changing the constitution to introduce an open term system for the presidency (Malunga 2012). Efforts to protect the rule of law at interstate level are exemplified by the work of the Civil Society Forum, the Southern African Litigation Centre (SALC), and the SADC Lawyers Association (SADC LA) who have been fighting to save the SADC Tribunal, which SADC member states suspended. The SADC LA and the Pan African Lawyers' Union (PALU) submitted a request to the African Human Rights Court to review the legality of limiting the mandate of the court from inter-state to intra-state matters (Freedom House 2012). The ruling on this is still awaited at the time of writing this paper.

Other important areas in which southern African CSOs have contributed are civic education, democratic citizenship, and human rights. The Namibian Institute for Democracy has championed civic education, as has the Zimbabwe Civic Education Trust (ZIMCET), which has, in addition, set up peace committees to spearhead peace and tolerance of divergent views. FBOs in Madagascar have been key in promoting civic education, democratic culture and conflict prevention. Against a background of gender injustice, and the HIV/AIDS pandemic, civil society in southern Africa has fought for the rights of women and children and for state-supported treatment of those infected. The Treatment Action Campaign in South Africa and the Southern African Network of AIDS Service Organisations (SANASO) have successfully lobbied governments to expand their responses to the pandemic (Godsäter and Söderbuam 2008). CSOs such as Women for Change in Zambia, Women and Law in Southern Africa (WILSA), Gender Links, and the Southern African Gender Protocol Alliance (SAGPA) have successfully lobbied their respective governments to implement health and development programmes that are gender sensitive and increase the share of women positions in strategic institutions. In the DRC, where over 200,000 women have been raped by armed men since 1998, CSOs such as Search for Common Ground (SFCG), are effectively working with the armed forces to prevent sexual and gender violence (SFCG 2012).

CSOs have also worked to enhance the credibility of the democratic electoral processes. In Malawi, around 30 years after majority rule, civil society used mass protests to press the first post-colonial President to expand their political freedoms (Immink et al. 2003), which resulted in the ushering of formal democracy and holding of multiparty elections in the mid-1990s. In Namibia the Council of Churches of Namibia, the Namibian Non-Governmental Organisations' Forum, and the Namibian Society of Human Rights have fostered public confidence in the electoral process by ensuring transparency (Kaapama et al. 2007), work that has also been undertaken by the Zimbabwe Electoral Support Network to a limited extent.

After a sustained history of human rights abuses southern African CSOs have contributed to national healing and reconciliation initiatives in the post-independence period. These initiatives include truth recovery, trauma counselling, victim support services, memorialisation and reparations, and the development of cultural models of reconciliation (Colvin 2007; NANGO 2010). Such CSOs include the Zimbabwe Human Rights NGO Forum and the Nairobi-based Agency for Cooperation and Research in Development (ACORD), which operates in 20 post-conflict African countries, including Mozambique, DRC, Tanzania, and Angola. In South Africa, the Institute for the Healing of Memories is one example of a number of organisations involved in this work.

Struggles to increase access to wealth

Economic justice remains a key theme for southern African civil society. Some of the focus has been on addressing relations between southern African states and the developed world. Civil society work is focusing on debt cancellation, better trade terms, and land resource rights. Within countries, civil society has been lobbying for more transparent budgeting and equitable distribution of national income; against corruption between government and the private sectors; and towards increased local ownership of means of production. Examples of CSOs driving this agenda are the Southern African Centre for Economic Justice (SACEJ), the Southern African People's Solidarity Network (SAPSN), the Zimbabwe Coalition on Debt and Development (ZIMCODD) (Godsäter and Söderbuam 2008), Zimbabwe's Affirmative Action Group (AAG) and Indigenous Business Women's Organisation (IBWO), and Namibia's Indigenous People's Business Council. The difficulty is that across the region, the benefits have tended to accrue to the elite.

In agriculture, genetic resources and land have become important economic justice issues. This explains why CSOs such as Participatory Ecological Land Use Management (PELUM) Association, the Eastern and Southern African Small Scale Farmers' Forum (ESAFF) and Biowatch are lobbying against corporate takeover of the agricultural production chain, where corporations create farmer dependence on them through terminator technology which ensures that farmers cannot recycle seed (Mzinga 2005). There is a new wave of land grabbing by foreign companies and governments (from the West and the East), which has partly been inspired by finite nature of oil and the demand for biofuel to replace it, and the surge in food prices which expose developed countries with low natural resources to food shortages (Palmer 2008; Hall 2011). When the government of Madagascar tried to lease over half its arable land for biofuel crop production to an Asian company, civil society responded with demonstrations in the country, contributing to the overthrow of the government (Ramiaramanana 2010). Biofuel-based land deals have been made in countries such as Zambia, DRC, and Angola (Hall 2011).

The Pan African Programme on Land and Resources Rights (PAPLRR) and LandNet Africa are two leading CSOs that have provided a framework for civil society engagement with land rights (Saruchera and Odhiambo 2004). However, PAPLRR has been forced to close because donors withdrew funding (Palmer 2008). The Land Campaign Coalition of Mozambique has participated in national land advocacy issues with some degree of success (Saruchera and Odhiambo 2004). In Zambia, Oxfam successfully lobbied for local people to secure land rights over land that had been previously the property of mining companies and absentee landlords in the Copperbelt (Palmer 2008).

Conclusion: major dilemmas faced by civil society in southern Africa

This paper shows the growth of a sector of civil society in southern Africa has been a response to the evolving needs and issues of society in the region – through colonialism, de-colonisation, and the honeymoon period of early independence into the present. Struggles to expand political freedoms, and increase access to wealth, have been at the heart of this growth. Although progress has been made in these areas, much work remains for civil society to do.

Over the past 20 years southern African CSOs have been formalised as a part of the funded development aid sector. In the process they have become highly dependent on international funding, which does not enable them to strategically tackle structural economic injustice, beyond short-term service delivery projects or social enterprise initiatives (Berkhout 2011; Civil Society @ Crossroads 2012). Donor dependence has sometimes led to a disconnect between CSOs and citizens, resulting in citizens' mass actions involving new and traditional forms of activism (CIVICUS 2011). This is illustrated by recent mass actions in Madagascar and Malawi. Southern African CSOs therefore face the challenge of contributing to fundamental and systemic societal change, which involves confronting powerful and dominant political and economic societal forces. This challenge is made more difficult as independent advocacy for reforming democratic governance "*does not have much resource support*" (Civil Society @ Crossroads 2012, 15).

References

Beauclerk, J., B. Pratt, and R. Judge. 2011. *Civil Society in Action: Global Case Studies in a Practice-based Framework.* Oxford: INTRAC.

Belay, N., A. Firmin, and C. N. Pegus. 2012. *State of Civil Society*. Johannesburg: CIVICUS.

Berkhout, R. 2011. "HIVOS and the Immaterial: Reflections on a Balancing Act." *CDRA: An Annual Digest for Practitioners of Development 2010/2011 – Investing in the Immaterial*, edited by S. Pothier, N. Dlamini and V. Weitz, 21–26. Cape Town: CDRA.

Bourne, R. 2011. *Catastrophe: What Went Wrong in Zimbabwe?* London: Zed Books.

Church and Civil Society Forum. 2010. *Church and Civil Society Submissions: Towards Development of a National Framework for National Healing, Integration and Reconciliation in Zimbabwe*. Harare: NANGO.

CIVICUS. 2011. *Bridging the Gaps: Citizens, Organisations and Dissociation. Civil Society Index Summary Report: 2008 – 2011*. Johannesburg: Civicus.

Civil Society @ Crossroads. 2012. *Civil Society @ Crossroads: Shifts, Challenges, Options?* New Delhi: PRIA.

Colvin, C. J. 2007. "Civil Society and Reconciliation in Southern Africa." *Development in Practice* 17 (3): 322–37.

Edwards, M. 2009. *The Challenges of Civil Society in Africa*. Dakar: TrustAfrica.

Freedom House. 2012. "African Human Rights Organizations Look to Court to Save SADC Tribunal. Accessed January 5, 2013. www.freedomhouse.org/article/african-human-rights-organizations-look-court-save-sadc-tribunal

Godsäter, A., and F. Söderbuam. 2008. "The Role of Civil Society in Regional Governance: The Case of Eastern and Southern Africa". Paper for GARNET Annual Conference Bordeaux, September 17–19, 2008.

Hall, R. 2011. "The Many Faces of the Investor Rush in Southern Africa: Towards a Typology of Commercial Land Deals." *Initiatives in Critical Agrarian Studies (ICAS) Review Paper Series 2*, 193–214. The Hague: International Institute of Social Studies (ISS).

Immink, B., S. Lembani, M. Ott, and C. Peters-Berries. 2003, eds. From Freedom to Empowerment: Ten Years of Democratization in Malawi. Balaka: Monfort Media.

Kaapama, P., L. Blaauw, B. Zaaruka, and E. Kaakunga. 2007. "Consolidating Democratic Governance in Southern Africa." *EISA Research Report 34*. Johannesburg: Electoral Institute for Sustainable Democracy in Africa.

Malunga, C. 2012. "Civil Society at Crossroads in Malawi: Where Do We Go after 20th July Mass Demonstrations." Unpublished paper, Cape Town: CDRA.

Mandaza, I. 1999. "Reconciliation and Social Justice in Southern Africa: The Zimbabwe Experience." In. *African Renaissance: The New Struggle*, edited by M. W. Makgoba, 77–90. Johannesburg: Mafube & Tafelberg.

Mlambo, A. S. 2009. "From the Second World War to UDI, 1940–1965." In *A History from Pre-Colonial Period to 2008*, edited by B. Raftopoulos and A. Mlambo, 75–114. Harare: Weaver Press.

NANGO. 2010. *Civil Society Constitutional Compendium: Conditions and Demands on Content and Process*. Harare: NANGO.

Orner, P., and A. Holmes, eds. 2011. *Don't Listen to What I Am About to Say: Voices from Zimbabwe*. Johannesburg: Jonathan Ball.

Palmer, R. 2008. "Land Reform in the Broader Context Of Southern Africa". Paper presented at a Workshop on the Land Reform from Below: Decentralised Land Reform in Southern Africa, April 22–23, in Johannesburg.

Pellicer, M., V. Ranchhod, M. Sarr, and E. Wegner. 2011. *Inequality Traps in South Africa: An Overview and Research Agenda*. Cape Town: SALDRU.

Perlas, N. 2000. *Shaping Globalization: Civil Society, Cultural Power and Threefolding*. Quezon City: Centre for Alternative Development Initiatives.

Ramiaramanana, D. 2010. "Impacts of Land Grabbing in Madagascar". Paper presented at a Regional Workshop on Commercialisation of Land and "Land Grabbing" in Southern Africa, University of Cape Town, March 24–25, in Cape Town.

Sachikonye, L. 2012. *Zimbabwe's Lost Decade: Politics, Development and Society*. Harare: Weaver Press.

Saruchera, M., and M. O. Odhiambo. 2004. "Civil Society and Social Movements: Advocacy for Land and Resource Rights in Africa." *Policy Brief: Debating Land Reform and Rural Development 12*. Bellville, South Africa: Institute for Poverty, Land and Agrarian Studies (PLAAS).

SFCG (Search for Common Ground). 2012. "Democratic Republic of Congo Programme. Accessed January 7, 2013. www.sfcg.org/programmes/drcongo/index.html.

Shivji, I. G. 2007. *Silences in NGO Discourse: The Role and Future of NGOs in Africa*. Oxford: Fahamu.

The authors

Mutizwa Mukute (corresponding author) is a research and development consultant, affiliated to Rhodes University, South Africa.

James Taylor is an Associate Practitioner at CDRA, South Africa. He has been involved in CDRA since shortly after its inception in 1987 – initially as a Board member, then becoming a member of staff in 1992. After moving to become co-director of CDRA with Allan Kaplan (the founder), eventually the executive director after Allan left in 2000, in 2011 James stepped out of the executive director position and reverted back to being a practitioner. With a background and training in community development work, James has worked as a fieldworker, facilitator/trainer, consultant, leader/manager, and occasional writer in the field of social development since 1976.

Struggles against systems that impoverish: South African civil society at the crossroads

James Taylor

This paper looks at the role of civil society in South Africa over the past 36 years. It views civil society as an integral part of a society that has undergone enormous change. It looks at civil society functioning as a part of a societal system that systematically impoverished the majority of its citizens. It traces the journey of people's struggles to survive and thrive and how these culminated in overthrowing an unjust political regime. It follows the journey through the political transition and looks at the point arrived at after the honeymoon period and the new struggle to make it work: arriving at a crossroads that, although a long way down the road, in the approach looks vaguely familiar. It ends by exploring a new way of thinking about the role of civil society that might contribute to finding ways of moving beyond systems that impoverish as a means of creating and concentrating wealth.

Luttes contre les systèmes qui appauvrissent : la société civile sud-africaine à la croisée des chemins

Cet article se penche sur le rôle de la société civile en Afrique du Sud au cours des 36 dernières années. Il considère la société civile comme faisant partie intégrante d'une société qui a subi d'énormes changements. Il examine la société civile comme fonctionnant dans le cadre d'un système sociétal qui a systématiquement appauvri la majorité de ses citoyens. Il retrace l'évolution des luttes des populations pour survivre et prospérer et la manière dont ces luttes ont fini par renverser un régime politique injuste. Il suit l'évolution à travers la transition politique et examine le point atteint après la période de « lune de miel » et la nouvelle lutte à mener pour obtenir de bons résultats : on se trouve à un carrefour qui, bien qu'il reste encore beaucoup de chemin à parcourir, a l'air vaguement familier sur le plan de l'approche. L'article se termine en examinant une nouvelle manière d'appréhender le rôle de la société civile, qui pourrait contribuer à trouver des moyens d'aller au-delà de systèmes qui appauvrissent afin de créer et de concentrer la richesse.

Luchas contra sistemas que empobrecen: la sociedad civil sudafricana en la encrucijada

El presente artículo examina el rol que ha tenido la sociedad civil de Sudáfrica durante los últimos 36 años. Su enfoque establece que la sociedad civil es una parte integrante de una sociedad que ha experimentado enormes cambios. En este sentido, la sociedad civil funciona como parte de un sistema social que ha empobrecido sistemáticamente a la mayoría de sus ciudadanos. El artículo describe la travesía que han debido realizar los pueblos durante sus luchas por sobrevivir y prosperar, analizando cómo dichas luchas llevaron al derrocamiento de un régimen político injusto. Continúa examinando esa travesía durante el proceso de transición política y en el periodo que tiene lugar tras el primer momento de "luna de miel", así como la siguiente lucha, realizada para que [el nuevo gobierno] tuviera éxito. Si bien el avance ha sido significativo, el acercamiento a la encrucijada no es una incógnita total. El artículo finaliza examinando una forma novedosa de pensar el rol de la sociedad civil, la cual podría contribuir a la búsqueda de sistemas que no produzcan empobrecimiento como mecanismo para crear y concentrar la riqueza.

Lutas contra sistemas que empobrecem: sociedade civil sul-africana na encruzilhada
Este artigo examina o papel da sociedade civil na África do Sul nos últimos 36 anos. Ele vê a sociedade civil como parte integral de uma sociedade que tem passado por enorme mudança. Ele vê a sociedade civil atuando como parte de um sistema de sociedade que sistematicamente empobreceu a maioria de seus cidadãos. Ele traça o trajeto de luta das pessoas para sobreviver e prosperar e como essas lutas culminaram no combate a um regime político injusto. Ele acompanha o trajeto desde a transição política e examina o ponto em que se chegou após o período de lua de mel e a nova luta para fazer isso funcionar: chegando em uma encruzilhada que, apesar do longo avanço, possui uma abordagem pouco familiar. O artigo termina explorando uma nova forma de se pensar o papel da sociedade civil que poderia contribuir para serem encontradas maneiras de se ultrapassar sistemas que empobrecem como forma de se criar e concentrar riqueza.

Introduction

> "Democracy is a process, not a static condition. It is becoming, rather than being. It can easily be lost, but never is fully won. Its essence is eternal struggle." (William H. Hastie, in Boyte and Skelton 1995, 2)

This paper looks at the role of civil society in South Africa over the past 36 years. It views civil society as an integral part of a society – seen as a living system[1] – that has undergone enormous change. It looks at civil society functioning as a part of a societal system that systematically impoverished the majority of its citizens. It traces the journey of people's struggles to survive and thrive and how these culminated in overthrowing an unjust political regime. It follows the journey through the political transition and looks at the point arrived at after the honeymoon period and the new struggle to make it work.

It traces a rich journey, arriving at a crossroads that, although a long way down the road, in the approach looks vaguely familiar. It ends by exploring a new way of thinking about the role of civil society that might contribute to finding ways of moving beyond systems that impoverish as a means of creating and concentrating wealth.

The story within this paper was started by bringing together a group of civil society activists in a workshop to share their experiences of how civil society has evolved in South Africa over the period. A biographical timeline was constructed, which was used to look behind the major events in civil society and the context to see patterns that give form and content to the story.

Second, the themes and insights arrived at through the workshop were used as the basis for a desk study expanding the themes of the story and linking them to existing writing and research.

The third part involved sharing the story verbally and in writing and engaging with audiences, both in South Africa and at international conferences. The telling and re-telling contributed significantly to shaping the latest version of a constantly evolving story, which this paper covers.

Systems that impoverish

> "He knew that the authorities were destroying him and his mind developed the metaphor of a machine. The machine was Robben Island. It was the apartheid system ... It was an apposite image which captured the ubiquitous, pernicious quality of the system that was bearing down on him." (Pogrund 1990, 303, describing the thought processes of Mangaliso Robert Sobukwe, founder of the Pan Africanist Congress, while a political prisoner on Robben Island)

South Africa has a history of colonisation and disenfranchisement, extraction and concentration of immense wealth and its accompanying systematic impoverishment. It has seen some of the world's most extreme examples of division. The apartheid system that institutionalised racial supremacy in law was finally toppled in 1994. After liberation the present fledgling democracy continues to grapple with one of the greatest gaps between the rich and the poor in the world (Bosch 2010).

To the oppressed, the inhumane institutions and agencies of the white regime were experienced as a powerful system. The term – "the system" – was a part of everyday language. When the security police appeared the warning would go out that the system had arrived. In describing what triggered the 1976 Soweto uprising, where students rose up against the enforced use of Afrikaans (the language of the minority white regime), one of its leaders explains the understanding of the system:

> "The disputed Afrikaans issue at junior schools such as Belle in Orlando West had resulted in many bitter, sometimes violent confrontations between pupils of schools and the 'system'. The 'system' was represented by everything associated with the Department of Education through to the hated policemen on the ground." (Mkhabela 2001, 44)

A more theoretical view of the interrelated system created between the state and business under apartheid is described as follows:

> "Capital accumulation would be based on the exploitation of a low-wage, highly controlled, expendable African work-force that was to be reproduced in a system of 'native reserves' at minimal cost to capital ... [after the 1948 election that brought the National Party to power] ... the National Party immediately set about implementing a rigorously codified racist project. Henceforth, race would become the definitive criterion for South Africans' access to privilege and opportunity, further restricting the social and economic mobility of black South Africans through a battery of legislative, administrative and other coercive measures." (Marais 2011, 9–15)

This story goes beyond seeing the system only in the enemy, to exploring the role of civil society as a foundational part of the society that impacts either positively or negatively on the lives of individuals and communities. It views society as a living system comprising of three interdependent realms – government, business, and civil society.

Civil society in South Africa

South African civil society at present is as diverse and contradictory (often uncivil) as the full spectrum of human need and interest. While formal organisations constitute only one small facet, the fact that there are 85,000 formally registered non-profit organisations (Funding Practice Alliance, correspondence with author, 2012) gives some idea of civil society's scope and vibrancy. While civil society organisations were crucial to creating and maintaining the inhuman apartheid system, it is the ongoing history of human struggle in pursuit of equality that provides the frame for this story.

Adopting a systems approach to exploring civil society focuses attention on the relations between the parts which connect them into a whole. The essential elements of civil society are ordinary citizens coming together in community around issues of common interest or need. There is a long history of the use of the term "civil society" and disagreement on a definition. Two recent definitions are helpful for my purpose as they both focus on the functions of civil society rather than on the more common debate around which organisations qualify as being in or out.

Ingrid Srinath offers the following understanding: *"... at its essence civil society is our species' response to the basic human need to come together in pursuit of common goals"* (CIVICUS 2011).

Beauclerk, Pratt, and Judge (2011) explore civil society as defined by the following five functions it fulfils in:

- helping generate the social basis for democracy
- promoting political accountability
- producing social trust, reciprocity, and networks
- supporting the rights of citizens and the concept of citizenship
- creating and promoting alternatives.

Srinath's definition cuts though the dramatic stories of the great leaders and the rich histories of political movements, through the volatile moments of uprising, protest, heroism, and sacrifice to underscore the role of ordinary citizens.

In South Africa, over time, this simple process of ordinary people coming together to pursue common goals towards meeting needs built towards the more elaborate and political functions of civil society outlined in the second definition. Through the social fabric of community the daily toils and strategies for survival, recreation, worship, study, and work, the relationships between the individual parts are forged into organisational wholes that in turn form parts of larger wholes.

The struggle

There are endless examples of people coming together to fight for their rights in communities all over the country. These slowly moulded into the popular base of what was referred to as "the struggle". Perhaps the zenith of this process was reached when 400 diverse organisations came together in forming the United Democratic Front (UDF) in 1983 with around three million members. "*The affiliates ranged from student groups to civic associations, overtly political organisations to sports clubs*" (Seekings 2000). The UDF later formalised their relationship with the Congress of South African Trade Unions (COSATU) forming the Mass Democratic Movement in 1989. "*The aim of the UDF, through uniting different organisations such as trade unions, was to make South Africa ungovernable – thereby crippling the apartheid system*" (South African History Online n.d.).

In South Africa the link between ordinary people coming together in communities and the civil society functions, as articulated by Beauclerk, Pratt, and Judge (2011), are clear, and we have seen them happen. The way this was done started changing in the 1980s. Prior to this the little support and protection provided to those organisations we now refer to as civil society organisations (CSOs) often came via the church, political, or trade union organisations. They had international links that created channels through which flowed relationships of solidarity and small amounts of funding, and "*were given full discretion to decide how the money was to be spent ... with donors adopting a highly accommodating attitude to local demands*" (Kihato 2001, 9).

But work in community or struggle organisations was driven more by commitment to the struggle than salaries or conditions of employment. Beyond finding whatever means possible to meet needs, the practice was shaped by the intentions contained in the slogan: "*conscientise, organise, mobilise*". The "*hostile environment ensured that state-civil society relations took an adversarial form*" (Habib 2003, 232) and organisations were held together more by the common cause brought about by a strong enemy than by sophisticated management skills.

While oppression increased in the 1980s, so did resistance. The civil society sector became a considerable force from within, making the country increasingly ungovernable.

> "Protest marches, boycotts, stayaways and strikes disrupted state structures in thousands of communities and schools. For weeks, or even months in some communities, the state was no longer in control of local governance." (Field 1988)

Meanwhile the political leadership of the liberation movement operated largely from exile.

The formalisation of civil society

Now, for the first time, representatives of international funding agencies started making direct contact with the smaller CSOs supporting the struggles and initiatives of the people. With funds came the first big shift in many organisations. They started formalising and the term NGO (non-governmental organisation) was increasingly used to describe them. New systems were called for, from employment contracts to funding contracts. Relationships and accountabilities were being formalised. Proposals and plans had to be submitted and reports written. (Even though not clear cut – from this point onwards the term NGO will be used to differentiate the more managerially constituted and funded support /service organisations from CSOs that we formerly described as "people's struggle" type organisations.)

By the late 1980s many of the NGOs battled to keep up with the growing demand to direct energy into building more efficient internal systems, procedures, and practices. But they learned quickly. The late 1980s and the 1990s were the heyday of the NGO sector in South Africa. The country became a priority for many international funding agencies. NGOs played a variety of roles in the final victory over the apartheid regime. Highly proficient organisations were in place to assist in the negotiations, to prepare for the first democratic elections in 1994, and to participate in attempts at national reconciliation and healing after generations of violence and inhumanity. NGOs played a significant role in the drafting of the policy and constitutional framework for the new government, and acted as a reservoir providing the skilled staff needed in the new government and in post-apartheid business (Dudouet 2007).

Transition and reconstruction

This shift from overthrowing a regime to being a part of building its replacement constituted a second major shift for both NGOs and CSOs. In the transition years from before the first democratic election in 1994 through the early years of liberation there was much work to be done. South Africa was in the limelight and international funding was flowing. The focus shifted from fighting for the rights of citizens to developing practical means of meeting human rights and needs. Now often highly skilled, NGOs were active in developing models of effective service in all areas of society. After decades of punishing and often life-threatening struggle those active in many CSOs felt free at last. They started to explore what it might mean to lead a normal life as a member of a democratic society. Many of the people's struggle organisations disbanded or took on new roles in the democratic processes.

The third chapter of change that has brought NGOs to the present crossroads has crept up over the past decade. Government settled into its new roles and responsibilities, with pressure growing to deliver on the people's expectations. NGOs' relationships with the state became more complex.

> "NGOs' somewhat hostile anti-government stance was replaced with a new identity (non-governmental), which ... reflects the relationship with government rather than a specific political or social orientation. NGOs and Community Based Organisations ... adopted the collective title of 'Non Profit Organisations' to reflect the(ir) depoliticised nature." (Mouton 2010, 69)

Many were committed to assisting in the delivery of government programmes.

But the honeymoon period was ending. The frustrated voices of people unable to access resources started growing again in protest against under-delivery (Desai 2002; Ballard et al. 2006). CSOs started supporting people in fighting for their rights and holding the government accountable. This resulted in increasing tension with established CSOs, particularly the better resourced NGOs receiving funds from international funders.

> "Criticism of government ... began to be labelled ... as counter-productive ... many organisations were in limbo, uncertain how to interact with a now legitimate and democratic state, and the people's voice in the body politic began to wane." (Naidoo 2010, 240).

Increasingly NGOs were challenged as not having the legitimacy of being mandated by the people, and accused of being agents of foreign agendas. "*Commercialisation and professionalisation ... blurred the non-profit/profit divide, and ... led to questions around the lines of accountability*" (Habib 2003, 234).

In addition, as the new millennium got underway, shifts slowly started happening in the flows of international development aid. South Africa started dropping as a priority as the focus moved to more vulnerable states and more recently emerging democracies. As South Africa became identified as an emerging market, expectations grew for the country to use its own wealth to close the growing gap between the rich and the poor.

Competition for funds increased along with funding conditions. Staff in CSOs "*complained that they felt more like bureaucratic aid administrators than development workers and that more time was spent on paperwork than development*" (Wallace et al. 2006, 165). It became more difficult to get funding for initiatives driven by the priorities identified by the organisations directly serving community level needs. Centrally planned initiatives ranging from the global Millennium Development Goals (MDGs) to national and local government priorities became increasingly crucial to setting development agendas. Monies flowed to those prepared and equipped to tender for short-term delivery contracts. CSOs advocating for social justice, and lobbying for policy changes and against mismanagement and non-delivery, started to find their relationships with government increasingly strained. Those winning delivery contracts from government grew ever closer (Borstein 2001).

The point we have reached

The bigger picture

Eighteen years on, South Africa is a very different place from the one it was in 1994. In arriving at the first democratic election, South Africa started throwing off its status of pariah state. Its previous geopolitical importance had been defined by the dynamics of the cold war. It was strategically positioned on the tip of Africa and was a significant source of mineral wealth. It was welcomed back into the international community at a time of triumphalism of the west, five years after the symbolic fall of the wall in 1989.

Internally, political power has shifted from a small exclusive racial minority to a vastly more inclusive majority. Attempting to address systemic inequality, the new government has invested heavily in housing, education, and provision of electricity and water. It was budgeted that in 2010, spending on government grants and social assistance would increase to around 3.5 per cent of GDP (Bosch 2010, 4), placing South Africa amongst the countries spending the largest proportion of their GDP on social welfare.

The axis of global economic power has moved to the East. South Africa's history over the past 300 years has been shaped by its strategic position on the trade route between Europe and the East. It is now being seen as a gateway into Africa – a continent of opportunity. As the influence of Europe declines, South Africa, a significant contributor to the past wealth and growth of Europe, now finds itself invited to become one of the powerful middle-income emerging market countries known as the BRICS (Brazil, Russia, India, China, and South Africa) countries.

Within the country the economic power remains in the hands of a small (largely unchanged) elite, with the gap between the rich and the poor one of the highest in the world and growing (Pellicer et al. 2011). Unemployment has devastated the aspirations of many who fought for a fairer dispensation. The scale of the problem ranges from the conservative official measure of around 25 per cent to an "*expanded*" measure hovering between 35 and 40 per cent (Marais 2011, 177).

Behind the political and economic realities of South Africa lies a global landscape that reveals cracks in the fundamental functioning of the underlying institutions. Both the neoliberal economic system, and the democratic political system that has complemented it in the previously dominant countries of the world, are in disarray. The countries that once led are increasingly being forced to face their inability to meet the needs and aspirations of their citizens and their inability to out-compete the emerging economic powers. The growing crisis is heightening the old struggle for power and the apportioning of blame between the state and the market. In South Africa, this dynamic is playing itself out in an emerging democracy urgently attempting to address the legacy of past injustice and inequality. In a recent public lecture celebrating the 100th Anniversary of the African National Congress (the ruling party) the Deputy President engaged with the issue in light of the commitment to fighting poverty, inequality, and unemployment:

> "For centuries, democracy has been serving as the best integument for the system of capitalism. However, developments within the Euro Zone have now dispensed with some of the key elements of democracy by prompting change of governments without going through elections … [this] call(s) into question concepts that have until now been universally accepted as axiomatic. These developments have global implications … [they] present a challenge to … intellectuals to figure out the practical implications for Africa in terms of democracy and economic development." (Motlanthe 2012)

Civil society in the picture

Civil society in South Africa remains vibrant and dynamic. People continue to organise around issues of common need and interest, fully utilising the democratic spaces and institutions of the new political dispensation. CSOs and NGOs have, and continue to, contribute substantially to all areas of societal striving.

The response to the HIV and AIDS pandemic was shaped by CSOs with the Treatment Action Campaign (an interesting CSO/NGO hybrid) at the forefront. It took on and won against both a misguided government and then – together with the same government – profiteering global drug corporations (Mbali 2004). In other areas of health, in education, land reform, housing, environmental degradation and climate change, disability, and endless others, civil society activity abounds. Civil society organisations continue to seek ways of addressing unmet needs in society. In the social justice and human rights fields the interpretation and implementation of policies, laws, and even the Constitution are constantly challenged and shaped by CSOs, as the South African parliament facilitates dialogue and public participation in the shaping of legislation and government policy.

Over the last year civil society has again acted to protect and promote the underpinnings of democracy. As the government prepared to pass into law a Protection of State Information Bill, which could potentially curb media freedom and "*cast a wide net in terms of potential offenders*" (Stone 2012), civil society showed again how it can come together from its disparate spheres of concern to rally together in the Right2Know Campaign. This coalition has already had substantial success in getting amendments adopted to what has been dubbed "the Secrecy Bill".

Today in South Africa there is a plethora of organisations, formal and less formal, large and small, that make up the intricate web of relationships that constitutes civil society. But much has changed. Many iconic and also lesser known NGOs are closing their doors or fighting to find ways of reinventing themselves to keep them open.[2] The formation of new CSOs and political parties are indications of citizens continuing to organise around present needs.

The flows of foreign funding are rapidly drying up (Kihato 2001). By 2008 when the international economic crisis struck it was clear that many traditional international funders were themselves in a similar predicament to those they funded. Many South African organisations had enjoyed longstanding partnerships with largely European-based international NGO (INGO) funding agencies. In the early years they had funded our organisations as acts of solidarity. As the years passed many of these agencies were identified as well-placed to channel larger amounts of official government overseas development aid.

The INGOs grew to be able to access and disburse these much larger amounts. But just as happened to their recipients, they too were slowly drawn into dependency relationships with their back-donors. The demands and conditions that came with the money resulted in the INGOs increasingly becoming agents of government policies, priorities, and practices largely driven by the MDGs. As the governments of Europe grapple with their own economic crises the INGOs are themselves approaching the crossroads.

Within the country there is a funding community that is growing slowly in its skill and in the amounts it distributes. These include: a National Development Agency; national and local government departments and ministries; corporate social investment funds; and private philanthropists and trusts. A centralised National Lottery distributes a percentage of their proceeds. Taken together, these constitute a substantial base of financial support. They have little of the experience of the international funding community but are becoming increasingly conscious of their role and the skills that it takes to play it well (Coalition on Civil Society Resource Mobilisation

2012). The less formally organised activities of civil society continue to be funded largely by voluntary contributions of time, money, and gifts in kind of community members.

There is a shift, both locally and internationally, to the increasing use of funding to turn NGOs into instruments to deliver the solutions, programmes, services, or priorities of those with the power of money (Dhunpath 2003–2005). Increasingly, in order to survive, NGOs have to tender for contracts to deliver pre-designed solutions. Results-based management and the growing monitoring and evaluation industry have become the dominant practices through which distant external and centralised command and control functions are exercised according to contractual obligations and agreements (Dlamini 2005/6).

This dynamic suggests some of the tension in the power dynamic between civil society and both the state and business, who share the ability to influence through funding. As the growing gap between the rich and poor is experienced as being fuelled by the greed and self interest of business and political leaders, civil society is again starting to shift its focus towards organising and mobilising at community level.

The question of whether we have come full circle in 18 years and if the time is coming to *go back to the trenches* is not only being heard in civil society, but there are signs of it being acted upon. Despite all government's attempts to compensate for the inability of the market to reverse the growing gap between the rich and the poor, community protests against the lack of service delivery by the state are common. Notwithstanding a significant decline from the previous year there was still an average of over 11 community protests per month in 2010, of which more than half were violent (Karamoko and Jain 2011).

At the crossroads

Meeting

The role and impact of NGOs is being questioned. Both those allocating the resources and those doing the work are recognising that the systemic causes of the social problems are not shifting. The familiar patterns of exclusion and frustration continue. Many of these organisations are facing closure if they cannot reinvent themselves in ways that attract the resources they require to continue their work. At the same time there is a re-emergence of vibrant civic/community-driven activist organisations and initiatives. The effectiveness of relationships between established NGOs and the more responsive organisations of citizens around issues of common concern are being questioned from both sides.

The notion of being at a crossroads is more relevant to those threatened with closure. They arrive at the crossroads, tired and convinced that together they have made a real difference. There is rich and obvious evidence of the contributions made to society through the societal functions identified by Beauclerk, Pratt, and Judge (2011) in their definition. And yet 18 years after liberation, despite the successes, there is little evidence of having created transformed systems that do not impoverish as they enrich.

The state and the market, too, are at the crossroads. At the time of writing the ruling party in South Africa is debating the need for a "*second transition*" (African National Congress 2012) at their policy conference. They too are grappling with the fact that their best efforts over 18 years have not resulted in the systemic changes they had envisaged and fought for. In Europe democracy – the very ideal that inspired so many struggles for liberation – is at the crossroads too. Globally, the capitalist economic system also finds itself in major crisis.

From a South African perspective human society collectively is at a crossroads. The challenge faced by the full range of civil society organisations is what alternatives they now have to offer. In South Africa there is a sense of teetering on a precipice. Behind there is the firm ground of going back to the trenches, of taking up the struggle again. Ahead is the unknown void of possibility, of creating and promoting alternative systems that support life in all its diversity.

Thinking

In South Africa we are sensitive to society's tendency to organise itself in ways that exploit and diminish human communities. We see clearly the parallels between environmental degradation and human exploitation. The world we are part of extracts human and natural resources in ways that impoverish the organisms and systems that have produced and nourished them.

The crossroads provides an opportunity for civil society to search for creative alternatives. The conviction is growing that the way we are organising ourselves is not working, and that harder work will further entrench the systems and the cycles by applying the same frameworks and analysis that have brought us to where we are. Shaped by past relationships, many NGO survival strategies are drawing them into closer relationships to the state and business. The more activist, political and civic driven CSOs risk remaining trapped in vicious cycles of alternating defensive and combative strategies that challenge but still fail to transform the system. To break these cycles a new consciousness is required.

> "We need a new consciousness. History is always responsive to a new consciousness ... and what defines a society is not how it overcomes its night but what it does with the long ever-after days of sunlight." (Okri 2012)

The challenge is to use as a source that fundamental impulse of people to flourish and develop to their fullest potential against forces that seek to control, to extract, and to limit: to engage the experience of those most excluded around a new consciousness and thinking born of the imperative of ecological sustainability.

Seeing patterns

The insights encountered at the crossroads prompts those intent on social transformation to experiment with applying a few rudimentary principles of an emerging new way of thinking known as systems thinking. This sees the world not as a machine but as a complex interconnected living organism. This way of thinking is often referred to as ecological as opposed to mechanistic. Its implications for how we see and contribute to the world are radical; we are only just starting to understand what this means for the way we will have to act in the future. It highlights the different approaches required when working in simple, complicated, or complex systems. Working with living human systems is by definition complex, resulting in it being impossible to control pre-determined outcomes through command and control management (Wheatley 2005).

Transformation in complex living systems will not be achieved through linear logic and short-term, results-based projects. While at the crossroads civil society has time to contemplate letting go of the linear cause and effect, input-output-outcome-impact thinking that results in command and control management. It can contemplate embracing more of the new ecological way of thinking that is equipped to inform meaningful engagement with complex adaptive self-managed systems.

Agents of change must develop a new understanding of how change takes place in complex adaptive systems. It is a radically different theory of how change happens to those that have shaped both the militaristic liberation struggles and the increasingly strategically planned NGO development activities.

> "Life's process for change is termed emergence, and it is how local efforts achieve global impact. In nature, change never happens as a result of top-down, preconceived strategic plans ... Change begins as local actions spring up simultaneously around the system. If these changes remain disconnected, nothing happens beyond each locale. However, when they become connected, local actions can emerge as a powerful influence at a more global or comprehensive level. (Global here means a larger scale, not necessarily the entire planet.)" (Wheatley 2005, 178)

It is characteristic of emergent change to appear suddenly and surprisingly. On reflecting on the struggle against the system in South Africa this understanding of change is helpful. There is much evidence of small local actions over long periods of time having minimal impact, then coming together with surprising power for change. The student uprising in Soweto is a case in point. Emergent phenomena are more powerful than the sum of their parts and possess capacities that are different from the local activities that resulted in them. Emergent change can result in positive or negative outcomes depending on the values and purpose around which local initiatives are organised.

Living systems need to be looked at over time as they progress through cycles and stages. Over the period considered in this paper the pattern of small initiatives born out of the human need to enrich life in community is evident. Processes of organising through complex relationships into larger organisations, networks, and institutions that connect smaller organisms and organs into larger societal systems have been experienced. A strong thread in the story is formed by patterns of separating and operating individually (differentiation) and occasionally coming together (integration) to meet higher order needs and challenge restrictive dominant systems.

Thinking this way the function of civil society is understood in the local relationships and actions between all citizens and their impact on shaping the institutions of governance and economics. The ecological view suggests that the form of the total system emerges from the values and purpose that shape the interactions between the constituent parts. The nature and quality of relationships within and between civil society, the polity, and the economy at local level is therefore significant in shaping society.

Nicanor Perlas presents a threefold systemic view of society. He describes the function of the systems simply:

> "The economy deals with harnessing nature to efficiently meet human needs ... [it] is that subsystem of society or the social world that deals with the production, distribution, and consumption of goods and services for the appropriate satisfaction of human needs." (Perlas 2000, 111)

Perlas describes the function of the political realm even more succinctly: "*Polity is the subsystem that deals with equity in all aspects of human relations*" (Perlas 2000, 111). Perlas views the function of civil society as foundational in shaping society through cultural creativity:

> "Culture is that realm which intuits the guiding ideas of society ... that gives identity and meaning ... that develops the full potential of individuals and enables them to be competent participants in the economy, political life, culture and society-at-large. Culture deals with the entire realm of ideas and includes worldviews, knowledge, meanings, symbols, identity, ethics, art, and spirituality." (Perlas 2000, 111)

Paul Hawken presents an interesting view of those organisations in civil society that act against the systems that impoverish:

> "we can imagine a collective movement that would protect, repair, and restore that organism's capacity to endure when threatened . . . like an immune system, which operates independently of an individual person's intent . . . the shared activity of hundreds of thousands of non-profit organisations . . . as humanity's immune response to toxins like political corruption, economic disease, and ecological degradation." (Hawken 2007, 141)

While these two ecological ways of describing the function of civil society differ in their emphasis, they both suggest that ordinary people coming together plays a formative role in shaping society.

Moving on

Looking back through this more ecological understanding, the emphasis shifts. Civil society's full contribution was much more than the struggle against the compelling horror of the values that drove apartheid. For instance, cultural activities such as songs, dance, and drama, were an important element in the struggle (Taylor 1997). The people who came together and organised themselves – the type of creative "*social inventors*" (Ndii 2008) that are found working in civil society the world over – were equally driven by values born of the need to flourish and to contribute towards shaping a society that maximises the potential that lives within it.

The strength and contribution of civil society is then to be found in the functional qualities of the alternative values and the purpose that the people organise around. Through their collaboration with like-minded organisations – were they "*to link up more self-consciously, within and across sectors . . . they can be expected to take strength from each other and encouragement to think beyond the boundaries of their separate assertions*" (Saul 2005, 223).

The recurrence of protests on the streets suggests that the immune system is detecting toxins in society again. It is deeply disturbing for people to see how difficult it has been to transform the economic and political realm into institutions shaped by the aspirational values enshrined in the Constitution. The tenacity with which values of greed and self-interest have managed to remain dominant is mobilising people again.

The work for organisations striving to transform society is to go to where the alternative values for sustainable society are being lived by ordinary people and, on getting there, to support and facilitate their attempts to organise and network around these values against forces driven by opposing values.

The challenge faced by transformative civil society as it meets business and the state at the crossroads is to bring its values and purpose more forcefully into the new relationships it is being drawn into. As it enters into the realm of social business and acts as delivery agent for the developmental state it must bring creative alternatives that connect to the struggles and strivings of ordinary people. But there is a danger of the essential function of civil society being lost as it is drawn into delivering on what are the responsibilities of business and government. Civil society's power and vibrancy are lost by its "*very participation . . . in . . . policy processes which rely on consensus-building and function within the framework of a market economy, which has tended to discourage a direct . . . approach to social issues, where CSOs are particularly effective*" (Papadakis 2006, vii–viii).

To deliver on its core function in the face of the needs of our time, all civil society organisations are being challenged to think and act differently. Resources must be directed towards struggles driven by the values of sustainability, diversity, and inclusion. Dominant ways of organising shaped by personal enrichment and greed must be countered. But there is a bigger challenge ahead. The future demands ecologically shaped ways of organising the economy and the polity that result in enriching living systems. Progressive parts of civil society could play their part by exploring and creating alternative ways of organising and relating at the local level.

Notes

1. For the present purposes, a system is simply defined as a set of interrelated elements (or sub-systems). A living system is open and self-organising with special characteristics of life that interact with and adapt to the environment.
2. See, for example, "An Emerging Funding Crisis for South African Civil Society" by Noxolo Kabane, *The Transformer*, December 9, 2010/Jan 2011, www.afesis.org.za/Local-Governance-Articles/an-emerging-funding-crisis-for-south-african-civil-societywww.afesis.org.za, and "The Great NGO Funding Crisis" by Rebecca Davis, *The Daily Maverick*, July 26, 2012. www.dailymaverick.co.za/article/2012-07-23-the-great-ngo-funding-crisis#

References

African National Congress. 2012. "The Second Transition? Building a National Democratic Society and the Balance of Forces in 2012." Accessed October 5, 2012. www.anc.org.za/docs/discus/2012/transition.pdf.

Ballard, R., A. Habib, and I. Valodia. 2006. "Voices of Protest: Social Movements in Post-Apartheid South Africa." Scottsville: UKZN Press.

Beauclerk, J., B. Pratt, and R. Judge. 2011. *Civil Society in Action: Global Case Studies in a Practice-Based Framework*. Oxford: INTRAC.

Borstein, L. 2001. "From Management Standards to Development Practice: A Study of the South African Aid Chain." CSDS Working paper 33. Durban: School of Development Studies and Centre for Civil Society, University of Natal.

Bosch, A., J. Rossouw, T. Claasens, and B. du Plessis. 2010. *A Second Look at Measuring Inequlaity in South Africa: A Modified Gini Coefficient*. Durban: University of KwaZulu Natal School of Development Studies.

Boyte, H. C., and N. Skelton. 1995. *Reinventing Citizenship: The Practice of Public Work*. Minneapolis: University of Minnesota. http://conservancy.umn.edu/bitstream/50414/1/6586.pdf.

Desai, A. 2002. *We Are the Poors: Community Struggles in Post-Apartheid South Africa*. New York: Monthly Review Press.

Capra, F. 2002. *The Hidden Connections – A Science for Sustainable Living*. London: HarperCollins.

CIVICUS. 2011. "Bridging the Gaps: Citizens, Organisations and Dissociation." Accessed August 10, 2012. www.civicus.org.

Coalition on Civil Society Resource Mobilisation. 2012. *Critical Perspectives on Sustainability of the South African Civil Society Sector*. Johannesburg: Jacana Media.

Dhunpath, R. 2003–2005. "It's All Businesslike Now." *The International Journal of Knowledge, Culture and Change Management* 3: 1109–24.

Dlamini, N. 2005/6. "Transparency of Process: Monitoring and Evaluation in Learning Organisations." Cape Town: CDRA.

Dudouet, V. 2007. "Surviving the Peace: Challenges of War-to-Peace Transitions for Civil Society Organisations." Berlin: Berghof Research Center for Constructive Conflict Management.

Field, S. 1988. "Politics and Consciousness in Cape Town Communities 1980–85." Accessed September 3, 2012. http://sahistory.org.za/archive/chapter-3-politics-and-consciousness-cape-town-communities-1980-1985-0.

Habib, A. 2003. "State-Civil Society Relations in Post-Apartheid South Africa." In *State of the Nation: South Africa 2003/04*, edited by J. Daniel, A. Habib and R. Southall, 232–234. Cape Town: HSRC Press.

Hawken, P. 2007. *Blessed Unrest: How the Largest Movement in the World Came into Being and Why No One Saw it Coming*. New York: Viking Penguin.

Karamoko, J., and H. Jain. 2011. "Community Protests in South Africa: Trends, Analysis and Explanations." Cape Town: MLGI.

Kihato, C. 2001. "Shifting Sands: The Relationship between Foreign Donors and South African Civil Society During and After Apartheid." Johannesburg: Centre for Policy Studies.

Marais, H. 2011. *South Africa Pushed to the Limit: The Political Economy of Change*. Cape Town: UCT Press.

Mbali, M. 2004. "The Treatment Action Campaign and the History of Rights-Based, Patient Driven HIV/Aids Activism in South Africa." Research Report No. 29. Berea: University of KwaZulu-Natal.

Mkhabela, S. 2001. *Open Earth and Black Roses: Remembering 16 June 1976*. Braamfontein: Skotaville Press.

Motlanthe, K. 2012. "Reflections on the Centenary of the African National Congress (ANC)." Accessed October 5. www.info.gov.za/speech/DynamicAction?pageid=461&sid=24704&tid=54782.

Mouton, C. 2010. "The History of Programme Evaluation in South Africa." Masters Thesis, University of Stellenbosch.

Naidoo, J. 2010. *Fighting for Justice: A Lifetime of Political and Social Activism*. Johannesburg: Picador Africa.

Ndii, D. 2008. Unnamed report presented at a seminar at the Sida Civil Society Center, Sweden, February 18–20.

Okri, B. 2012. "Steve Biko Memorial Lecture: Biko and the Tough Alchemy of Africa." Accessed September 13. www.uct.ac.za/news/lectures/stevebiko.

Papadakis, K. 2006. "Civil Society, Participatory Governance and Decent Work: The Case of South Africa." *ILO Research Series* 112. Geneva: International Institute for Labour Studies.

Pellicer, M., V. Ranchhod, M. Sarr, and E. Wegner. 2011. "Inequality Traps in South Africa: An overview and research agenda." Cape Town: SALDRU.

Perlas, N. 2000. *Shaping Globalization: Civil Society, Cultural Power and Threefolding*. Quezon City, Philippines: Centre for Alternative Development Initiatives.

Pogrund, B. 1990. *How Can a Man Die Better: The Life of Robert Sobukwe*. Johannesburg: Jonathan Ball.

Saul, J. 2005. *The Next Liberation Struggle: Capitalism, Socialism and Democracy in Southern Africa*. Scottsville: UKZN Press.

Seekings, J. 2000. *The UDF: A History of the United Democratic Front 1983–91*. Cape Town: David Philip.

South African History Online. n.d. "Mass Democratic Movement (MDM) begins their Defiance Campaign." Accessed October 5, 2012. http://sahistory.org.za/dated-event/mass-democratic-movement-mdm-begins-their-defiance-campaign

Stone, C. 2012. "The 'Secrecy Bill': Taking Stock." Accessed October 5. www.r2k.org.za/2012/06/28/the-secrecy-bill-taking-stock-caroline-stone

Taylor, V. 1997. *Social Mobilisation: Lessons from the Mass Democratic Movement*. Cape Town: SADEP.

Wallace, T., L. Bornstein, and J. Chapman. 2006. *The Aid Chain: Coercion and Commitment in Development NGOs*. Rugby, UK: ITDG.

Wheatley, M. 2005. *Finding Our Way: Leadership For an Uncertain Time*. San Francisco: Berrett-Kohler.

The author

James Taylor is an Associate Practitioner at CDRA, South Africa. He has been involved in CDRA since shortly after its inception in 1987 – initially as a Board member then becoming a member of staff in 1992. After moving to become co-director of CDRA with Allan Kaplan (the founder), eventually the executive director after Allan left in 2000, in 2011 James stepped out of the executive director position and reverted back to being a practitioner. With a background and training in community development work, James has worked as a fieldworker, facilitator/trainer, consultant, leader/manager, and occasional writer in the field of social development since 1976.

Kampala city traders (KACITA) strike for action

Jackline Kabahinda

The Kampala city traders' strike in 2012 provides a vivid representation of the business community presenting and representing the interests of the citizens amidst a financial crisis in Uganda. The strike came as a shock to the powerful central government of Uganda. In the context of the strike, the crossroads for civil society is related to the capacities and relationships to coordinate the wider civil society to champion citizen's issues and well-being.

Les négociants urbains de Kampala (KACITA) font grève pour obtenir des actions
La grève des négociants urbains de Kampala en 2012 représente de manière parlante la décision par la communauté des affaires de présenter et de représenter les intérêts des citoyens dans un contexte de crise financière en Ouganda. La grève a été un choc pour le puissant gouvernement central de l'Ouganda. Dans le contexte de la grève, le carrefour auquel se trouve la société civile est lié aux capacités et aux relations pour coordonner la société civile dans son ensemble afin de défendre les questions concernant les citoyens et le bien-être de ces derniers.

La huelga de los comerciantes de la ciudad de Kampala (KACITA) para promover acciones
En 2012, la huelga de los comerciantes de la ciudad de Kampala resultó ser un ejemplo vívido acerca de cómo la comunidad comerciante presentó y representó los intereses ciudadanos en medio de una crisis financiera en Uganda. La huelga sacudió al poderoso gobierno central del país. En el contexto de la misma, la encrucijada de la sociedad civil se centró en las capacidades y en las relaciones que se pusieron en juego para coordinar a todos los sectores que la conforman, a fin de que éstos impulsaran soluciones a las problemáticas y el bienestar de los ciudadanos.

Comerciantes da cidade de Kampala (KACITA) fazem greve por ação
A greve dos comerciantes da cidade de Kampala em 2012 é uma representação vívida da comunidade empresarial apresentando e representando os interesses dos cidadãos em meio a uma crise financeira em Uganda. A greve veio como um choque para o poderoso governo central de Uganda. No contexto da greve, as encruzilhadas da sociedade civil estão ligadas à capacidade e às relações para se coordenar a sociedade civil mais geral para promover as questões e bem-estar dos cidadãos.

Introduction

> "Civil society minus forces reserved for action and reaction is irrelevant."
> (Interview with Issa Sekitto, spokesperson for Kampala City Traders Association, Kampala, May 25 2012)

The Kampala city traders' strike in 2012 provides a vivid representation of the business community presenting and representing the interests of the citizens amidst a financial crisis in Uganda. The strike came as a shock to the powerful central government of Uganda.

In a bid to influence the harsh economic and business environment in Uganda, mainly stemming from high bank interest rates that the traders found intolerable, the traders' association organised a strike that grew from one day to several days and later gained support from the general public. While the strike succeeded in bringing the Ugandan government to the table to talk with representatives of the association, it is less clear that the strike succeeded in compelling anything other than cosmetic changes in government policy. However, the strike reflects a growing commitment from middle-class constituents to challenge the status quo and government policy, and to undertake advocacy with the Government, thereby reflecting the emerging activism among citizens – in this case the traders rather than the poor and marginalised groups.

To begin, this paper briefly conceptualises civil society in general and explores the concept more specifically towards contextualising "the traders strike". It then goes on to describe the details of the strike, and draw up some implications for policy and practice.

Conceptualising civil society

The term civil society has had astonishing trajectories and inroads – be it in politics, academia, and local community forums. Perhaps one of the important outlooks in relation to this discussion is from John Locke and Adam Ferguson, who laid the foundation for civil society use in modern societies – seeing civil society as a society ruled by a social contract among free men to establish a lawful order. A contrasting conception by Montesquieu and Tocqueville stresses the relevancy of associations as means by citizens to limit the powers of the state and to develop values. Influenced by experiences of the French revolution, Hegel conceptualises civil society by including commercial activities as an important component of modern society, under the supervision and guidance of a strong state (Kruip and Reifeld 2007).

For the purposes of the discussion here however, it's useful to look at civil society as a place where people come together to advance their common interests, essential for the functioning of inclusive communities, democratic political systems, and stable societies as a whole.

Civil society is usually spoken about in relation to two other concepts; non-governmental organisations (NGOs) and civil society organisations (CSOs). I will briefly explore these now.

NGOs, usually referred to as non-state actors, legally constitute private, voluntary groupings or associations, organisations with no participation and representation by government; "the independent sector", and including volunteer sector, civil society organisations, grassroots organisations, transnational social movement organisations, private voluntary organisations, and self-help organisations.

NGOs are constituted by diverse groups of individuals with diverse objectives. Some are indigenous while others are international or foreign-based. NGOs differ in their membership or constituencies; geographical dispersion; motivation and values; leadership, objectives, role(s), methods of work, funding, and capacity. They are mostly engaged in service delivery and are spread throughout in the country. Unlike their local counterparts, they are more secure in their funding and have sufficient capacities to engage the government in policy processes. They are also more visible in engaging on policy issues.

On the other hand, indigenous NGOs are largely characterised by local membership – one that is predominantly urban, and preoccupied with service delivery roles as opposed to advocacy work.

Civil society organisations (CSOs) in Uganda are categorised on the basis of their membership or constituencies, geographical dispersion, motivations and values, and their roles, activities, and functions. However, given the wide diversity inherent in the nature and composition of CSOs in Uganda, it is difficult to classify them in a coherent manner. A categorisation that reflects their constituencies as well as their main activities or functions, is used to guide better understanding of their societal role.

Community based organisations (CBOs) are a less formal and more spontaneous category of civil society constituted by informal mutual self-help, developmental organisations based in both urban and rural areas. CBOs are scattered all over the country and are usually unregistered and deal with local village or sub-county level issues. They constitute the largest part of civil society. They include informal savings and credit groups, and burial associations. CBOs aim to address the needs of the people that are far from the reach of the state. There is thus a growing interaction and dialogue between CSOs, citizens, and government structures in Uganda, particularly in the policy formulation and implementation processes.

Background of civil society in Uganda

In the period before the country attained its independence from Britain, some organisations existed mainly to resist colonial oppression and exploitation. Mamdani (1983) contends that prior to Uganda's independence the country had two strands of CSOs: elitist groups demanding equal opportunities with colonialists (for education, employment, etc.) on the one hand, and militant groups often comprising the peasantry and workers, demanding higher commodity prices and political independence, on the other. He further argues that the colonial state provided little or no space for civil society activism and it "*sought to control the evolution, content and impact of associational life in Uganda… in order to prevent drastic challenges to the hegemony of the colonial state*" (Oloka-Onyango 2000a, 4).

It should be noted that the emergence of formal civil society associations representing collective interests in Uganda first took place under colonial rule. The early missionaries and the religion they brought with are recognised as significant, and key towards institutionalising civil society.

During colonialism, welfare and charitable organisations were often founded under the auspices of the church, such as the Red Cross Society. These were distinct from the membership-based organisations in that they acted as philanthropic intermediaries providing welfare services to the poor. In this role they were the early precursors of the non-governmental organisations that rose to prominence in later years. Charitable organisations were founded under Christian values and in so doing promoted Christian ideals of caring for the needy. There is no doubt that the Church and similar institutions have contributed positively to the institutionalisation and transformation of Ugandan society, for in society they belong as citizens.

Religion, religious values, and religious institutions have influenced societal values and culture through the assumed role of service delivery, particularly through collaborations and partnerships with civil society organisations. Therefore the concept of interdependency between the church and civil society is visible in Uganda.

The Protestant Church, the Catholic Church, the Islamic faith, and other religious/faith groups often acting separately have contributed generally to Uganda's development and the shaping of institutionalised civil society. Significant to mention is their historical role in shaping the educational institutions, thereby liberating humanity and in the process facilitating and enabling development by unleashing human potential.

Religious or faith-based CSOs have had a strong role in the provision of social services – particularly in the areas of health, education and community development. They have also been prominent in influencing policy positions over a wide range of issues, such as reproductive health (the debate over abortion and the use of contraceptives) and polygamy. For instance, objections by the Catholic Church to the use of condoms as a means to prevent the spread of HIV/AIDS resulted in the imposition of restrictions on the nature and extent of advertising of condoms on state-controlled media. The Catholic Church was influential in defining the parameters of the recognition of the right to abortion during the Constitutional debate. Similarly, on the issue of polygamous marriages, vehement objections were raised by members of the Islamic faith to proposals under a new Domestic Relations law aimed at restricting the number of wives one can marry to two and not four, as is the position under Islamic law. Eventually, the proposals were withdrawn (Oloka-Onyango 2000a). The Church of Uganda has also contributed to the debate on the Anti-Homosexuality Bill,[1] expressing concerns with the bill, and recommending that the bill be amended to reflect that homosexuality, the practice or the promotion of homosexual relations, is not adopted as a human right.

Therefore, depending on the issue present in society, the Church has adopted different positions ranging from service provision, to voicing societal concerns, and influencing policy and policy decisions. Eminent also is a definite relationship between the political community, the church and significantly with the societal community. It therefore remains to be seen how the church in its self-defined role as an agent of Christ through evangelization could reconcile the fragile and the broken whole to which the CSOs, the private sector, communities, and individuals are gifted. This is also in consideration that the political community and the church are supposed to be mutually independent and autonomous – yet each serving the personal and the social human being.

Further categorisations of civil society forms in Uganda include:

Mass-based membership organisations, formed in the early 1990s to promote the economic and social interests of peasants and workers, and included peasant-based cooperatives formed in the inter-war years to resist the monopolisation of trade in agricultural products by the colonial state and immigrant communities from Europe and Asia. Cooperatives and trade unions were also formed in this period to address labour-related issues including low wages and poor working conditions. It was these trade unions that organised the general strike of 1945, leading to an increase in wages and an improvement in workplace conditions (Bazaara and Barya 1999; Bazaara 2000).

This category is more representative, is established by law and concentrates on issues that directly affect their membership. They are well organised in the country partly because they represent the educated and professional elite (Bazaara and Barya 1999).

Elite-led membership organisations, formed during the colonial period by middle-class Africans aggrieved at colonial policies. These included organisations such as the Young Men of Buganda, the Young Men of Toro, and the Uganda African Welfare Association (Bazaara and Barya 1999). In addition, some women's organisations, including the Uganda Council of Women (UCW), were founded at this time. UCW members rallied around collective women's rights issues such as citizenship, civic education, and voting rights, and they actively pushed for increased female participation in national politics (Tamale 1999).

The media. In 1918, the first newspaper, the *Uganda Herald*, was started. It published critical articles on the colonial regime and addressed concerns on land tenure; trade by Asians and Europeans; prices of agricultural products, and grappled with issues of colonial governance, exposing the exploitative and oppressive nature of its system.

To date, the media represents different social and political actors with different interests and agendas. In this sense, we can identify specific media for civil society and media for political parties. Several media houses have exposed and discussed cases of human rights violations and corruption. Despite the current ruling party's (The National Resistance Movement) efforts to regulate and control the media in Uganda, the Ugandan media continues to be one of the strongest advocates for regime accountability and openness, particularly those owned privately like the Nation Media Group.

Advocacy groups. The most prominent advocacy CSOs in the country work in the field of human rights – mostly advancing civil and political rights. Despite their mandate, however, human rights groups in Uganda are reserved about raising the more difficult and political human rights issues, and their mobilisation seems to be limited to a small network of people, usually in the urban setting.

Umbrella organisations and networks. Several NGOs have attempted to develop linkages and networks to improve their capacities, coordination, and to enhance their influence on the state. Such networks exist at the national, regional, district and sub county levels – the second lowest level of administration. They are mostly engaged in advocacy for improved service delivery at the local level of governance. Apparently, most donors prefer to relate to networks instead of individual organisations and promote their creation.

Ethnic-based organisations/cultural groups. Most of the cultural groups that exist in the country have historical and traditional linkages with kingdoms and ethnic groups. Their primary interests lie in the promotion of traditional leadership, language, music, dance, and drama. Some of the most influential and active cultural institutions/associations are those that were restored in 1993, particularly in Buganda, Bunyoro, Busoga, Tooro, and Ankole.

The Buganda Kingdom has been at the forefront of demands to return the country to federalism as a basis of governance. The kingdom has played and continues to play a pivotal role in the politics of the country. This is mainly because of its large population, relative economic prosperity, centrality in the country, and the popularity of the Kabaka. The kingdom of Buganda has engaged the government on issues such as land reform and governance. This has led to the Kingdom not being on good terms with the ruling government.

The Kingdoms of Buganda, Busoga, Bunyoro, and Tooro have established strong administrations with governance structures that appear parallel to those of the central government. They use these to mobilise their population for social and cultural causes. However, their reach and impact is limited by the fact that the traditional leaders are barred from engaging in active politics, and that they have no access to public funds to finance their activities.

It's noteworthy to mention that in the period after Uganda's independence in 1962 and especially after the abrogation of the country's first Constitution in 1967, the existence and operations of civil society was further abridged by the dictatorship regimes of leaders like Idi Amin and Obote. During Amin's regime, there was total breakdown of law and order as well as atrocious violations of fundamental human rights and freedoms. Press freedom[2] and associational life in the country was very much weakened, and the few traditional civil actors such as trade unions[3] and cooperative societies were largely inconsequential in impacting on political, socioeconomic, and other processes within the government's domain.[4] Oloka-Onyango points out that, for the few CSOs that remained in this period, issues of constitutionalism in human rights in particular were not prominent on their agenda (Oloka-Onyango 2000b, 3).

It is also important to note the developments in the country in the 1990s that affected traditional membership-based CSOs, particularly trade unions and cooperatives. Among these developments were the introduction and impact of liberal economic policies – Structural Adjustment Programmes (SAPs) – with its attendant policies of downsizing, privatisation, and restructuring. The developments contrast with the reduction in the number of public enterprises, as the power of many working class organisations associated with them reduced. This development led to another category of civil society: *development oriented and service delivery organisations.*

Their objective is to provide services to communities, mostly in the fields of health, education, legal expertise, social welfare, or to boost income generation and poverty alleviation. Several are largely non-membership-based and serve as intermediaries providing assistance to the poor, rather than as popular bodies representing the aspirations of farmers, workers and marginalised social groups. This category has a close relationship with the SAPs, and the rise of development and service organisations is particularly related to the removal of government subsidies to education and health and the introduction of cost sharing, which meant that the poor could not afford to pay for those services. Many organisations were formed to fill this vacuum, and became a source of employment for the retrenched civil servants. The relationship between this type of civil society and the state has been harmonious and cooperative – with the government sub-contracting them to deliver development projects.

Civil society and the private sector

Uganda's civil society and its private counterpart rarely meet. The private sector is yet to identify or associate with the causes of CSOs and has therefore not been visibly supportive. This was evident in the case of KACITA, even though the issues that led to the strike were crosscutting and had an overall impact on the general populace and citizens. This scenario therefore presents a situation of a lack of coordination and identity among the sector and consequently the role that the sector exists to, and should, play.

Current civil society operating environment

The current laws governing the sector, particularly NGOs, are said to be highly restrictive, allowing interference in NGO activities. NGOs are subjected to excessive bureaucratic control through mandatory registration, which is subject to broad executive discretion. Additionally, government officials are granted wide-ranging powers to limit the scope of NGO activities.

The Ugandan Parliament is currently also considering the Public Order Management Bill 2011, which will place a number of restrictions on the freedom of assembly, including the requirement to inform the police seven days in advance of holding public meetings. The bill also gives the police broad powers to control public assemblies, empowering them to 'direct' the conduct of public meetings and interfere with demonstrations on vaguely defined grounds.

According to Hassan Shire, Executive Director of East and Horn of Africa Human Rights Defenders (EHAHRDP): "*The situation in Uganda is rapidly deteriorating as civil society members are being intimidated in an attempt to stop them airing their views.*" Accordingly "*respect for freedom of expression, association and assembly are being restricted, yet they are all essential foundations of a democratic and rights-respecting society*" (URN 2012).

Furthermore, the operations of NGOs are regulated by the NGO Act, which requires organisations to register with the government's NGO Board, managed by the Minister of Internal Affairs. Members of Uganda's intelligence services sit on the board to monitor civil society activity. In this way, organisations have argued that they are being treated as possible national security threats.

The NGO Act as amended in 2006 is viewed as restrictive to the operations of NGOs due to lengthy and convoluted registration requirements and confusing procedures that groups are expected to comply with in order to receive permission to conduct research. In April 2009, eight organisations filed a challenge to the act before the Constitutional Court, arguing that some provisions are inconsistent with the Constitution, the International Covenant on Civil and Political Rights, and the East African Community Treaty. The case is yet to be scheduled for a hearing.

Setting the context

As Uganda prepared to celebrate 50 years of independence in 2012, the country faced many daunting challenges. Many of these challenges are said to be reflected in how the people, the society feels and acts. As remarked by one of the presenters at the CSO Fair held in Kampala in June 2012,[5] "*If you want to know how angry the Ugandan society is, just observe how we drive on the roads or watch how people behave in the Bukedde TV news 'Agataliiko Nfuufu'*" (interview with Sarah Nambuya, Kampala, March 2012) – literally translated as the "news without dust".

According to Sarah Nambuya, a development practitioner and member of APODEA (Association for Practicing Organisational Development in East Africa), many of the presentations and discussions at the CSO Fair depicted a number of crossroad issues for Uganda. She notes that largely, there is a sense of *helplessness* and *uncertainty*, although certain groups of people have started promoting greater awareness and transformational action in these crossroads situations. One of the many questions that still remain therefore is how prepared Ugandan society is to organise and coordinate efforts towards addressing the current challenges and opportunities that society faces.

2011 in particular posed unique challenges and opportunities to Ugandan citizens. Ugandans started the year with an election fever, where most were not sure what the immediate aftermath of the election would be like. After the 16 February 2011 elections, there were waves of protests, such as "walk to work" and teachers' strikes, to name but two.[6]

The volatility of the Ugandan shilling on the foreign exchange market did not help matters either. All these events impacted on Ugandan citizens in one way or another. The KACITA (Kampala City Traders Association) strike was one of those events whose significance was felt by the general populace, although it was private sector-led activity in pursuit of their interests.

In June 2011, city traders through KACITA had presented the Ugandan government with a 12-point issues paper which included, among others, the high prices of commodities, including fuel, which had a direct impact on consumption; the Trading License Act – which they wanted the government to review as well as the policy on foreigners doing petty trade in Uganda; and poor infrastructure, particularly roads and electricity. The issues were documented by KACITA and presented to the Ugandan government for action. KACITA and the government forged a way forward on most of the issues.

Given the situation in relation to the foreign exchange, the traders through KACITA proposed to the government that it intervened and fixed the exchange rate for at least three months. This negotiation was based on loses the traders were experiencing from the importation of goods, particularly from China where the dollar was depreciating. The rising inflation situation was aggravated by the Bank of Uganda's intervention to increase the Central Bank Rate (CBR). This had a particularly negative effect, for instance: a trader was expected to deposit US$120 on an existing loan; however with the increased CBR, the deposit increased to US$200, affecting the mortgaging in terms of increased rents, purchases of consumables for families, and so on.

Other consequences were immediately felt in the overall economy and lives of individuals and families. Inflation rate shot up from 10 per cent to 27 per cent. The exchange rate between the Uganda Shilling (USh) and the US$ went up by 16 per cent in favour of the dollar. The price of bread rose from US$0.72 per kilo to US$1.40. The cost of sugar increased sharply from USh 0.72 to between US$3.20 and US$3.60 in some parts of the country. All this happened within a span of three months in the last quarter of 2011. The impact was devastating across the board. Importers could hardly afford to buy foreign currency. The average citizen did not only have to carry the burden of inflation, but also, as its consequence, was confronted with harsh limited access to basic consumption needs.

Sequence of KACITA events 2011–2012

14 October 2011: KACITA Chairman, Everest Kayonda holds a press conference in response to the temporary closure of Kinyara and Kakira sugar factories, which led to an increase in sugar prices.

17 December 2011: KACITA threatens to petition the speaker of Parliament over the prevailing high interest rates by commercial banks, which was perceived to be dangerous for the borrower.

11 January 2012: Kampala traders' shut down shops following government's failure to intervene on the high interest rates charged by commercial banks on old loans to their customers.

12 January 2012: KACITA continue to close their shops for a second day, demonstrating against high interest rates on old loans. Most shops in the capital city, Kampala, remained closed as traders met President Museveni over the matter.

13 January 2012: KACITA leaders address traders on Luwum Street about what the President had proposed. The rally ended as most traders walked away demanding a more practical intervention. Police are deployed to keep law and order. A three-day strike starts and paralyses business in Kampala.

The Kampala city traders' strike

The January 2012 strike

Through KACITA, the traders decided to petition the President, asserting that old loans should stay at the rate with which they had been acquired. KACITA also selected a group of parliamentarians to convince the government and the Central Bank to explore other options. This latter effort also came to no avail.

After failing to reach an agreement, the traders announced a three-day strike. The government responded by stating that the strike was uncalled for and inappropriate, given that it would endanger the economy by scaring away investors. Although the strike was planned for three days, it extended to four days pending a government decision and communication on the issues raised by the traders.

Although the president had promised the traders' representatives a response to their issues within a week after the strike was called, there has been no official response to date, although the Central Bank Rate (CBR) was reduced by 2 per cent in mid-2012. KACITA officials expressed disappointment about the government's foot dragging in solving issues affecting the country. In this, and other upcoming activities, KACITA plans to mobilise countrywide to put enough pressure on the government to act more positively.

According to interviews with KACITA officials, the present and biggest challenge is that civil society in Uganda to date tends to be formal in its engagements with government. Accordingly, very few civil society organisations made a statement regarding the issues that were affecting almost everybody in the country, including the salary earners.

What drives citizen concerns in Uganda?

During a reflection meeting, organisational development practitioners in Uganda highlighted governance, and particularly leadership, as the key areas currently impacting on citizen welfare in Uganda.[7] It was noted that leaders at various levels of governance tend to politicise all issues affecting citizens, without putting in place systems or mechanisms for addressing their concerns. A number of citizen actions to address their needs under different circumstances illustrate this point, for instance, the stories of Kanungu and Pader in Box 1.

Box 1: Community and citizen-led activism

Kanungu residents re-build a bridge after delayed efforts by Government
According to the *Monitor* newspaper on Thursday 14 June 2012, at least 1,000 residents from Kanungu District started rebuilding a collapsed bridge after a meeting between President Museveni and district leaders could not provide a quick solution. Local people from Kambuga and Nyakagyeme sub-counties in Kanungu and Rukungiri districts felled trees and logs to the site of the former Mitaano Bridge on River Mirera. According to the area MP Chris Baryomunsi (Kinkiizi East) who apparently joined the residents in the task, "*we have received several complaints from the people in these two sub-counties that they find it difficult to cross the river to go to their gardens and visit their relatives*".

Community led citizen activism – the case of Pader
In Pader, one of the war-affected districts in Northern Uganda, the community mobilised themselves following a situation where the government had used the community land to plant electricity poles to transfer power to other parts and towns, while bypassing this particular community. Some members of this community first responded by cutting down a few poles; and in response the government imprisoned these individuals. Upon seeing this action by the government, and realising that the government did not take the community action seriously, more and more members joined in the cutting down of more electricity poles. The community was persistent and consistent with the action of cutting the electricity poles until the government connected this particular village with electricity. Those who were in prison were released unconditionally.

What emerges from the stories in Box 1 is that, in the midst of the perceived *helplessness* in sections of society with regard to issues affecting citizens, there is, however, some reasonable action and voice by the affected local communities, especially with regard to demands for a responsive government and a functional state.

In this sense, the traders' strike – a private sector strike – and its effects on the everyday populace gave it significance. The impact of the strike on the availability of essential goods gave the strike political significance, and it is likely that it would have passed unnoticed if it had not had this impact.

The significance of the KACITA strike

The KACITA strike is significant in a sense that we see the business community (private sector) presenting and representing the interests of the citizens, in a situation where the citizens feel helpless and uncertain about the future. The realisation that people can react through a strike was a shock to some actors and the progress made following the strike interventions led to the recognition that the strikers (traders and citizens) have the power.

It is also a story where the vibrancy of the media is visible – the strike was highly profiled by the Ugandan press and media houses, particularly the Nation Television (NTV) owned privately by the Aga Khan Foundation, and thus gained some prominence. Some towns (Masaka, Jinja, Mbale) joined in the strike having watched what was happening in Kampala. The strike however, revealed familiar lessons of how the civil society ought to overcome its own problems within the Uganda political system.

The KACITA engagements and strike confirm an emerging scenario, where most if not all strikes have ended with talks or seeking an audience with the President – who in turn has made promises that have not been followed up. To Ugandans, this is not a surprising phenomenon, as it has happened several times over the years – following teachers, students, and nurses' strikes prior to the traders' actions. Thus, the central role of the President as a last resort arbiter comes in, however, at a high price.

Overall, there were also political ramifications, including further protests related to political manipulations of the economy to benefit the government's political and economic interests.

Implications for practice

The KACITA strike confirms the reality that civil society actions are largely urban based, and in most cases lack participation and contribution from the different communities that form the wider civil society. With minimal representation from rural areas where the majority of the people reside, demonstrations and strikes like that of KACITA are likely not to make significant impact. It is therefore important to note that the success of any demonstration or strike depends on broad acceptance in the distribution of social power.

The organisations that promote social power must themselves be connected to, and accountable to the society. CSOs thus ought to build certain capacities beyond the mobilising power and skills to succeed in engaging the government and the demands that they set out. Thus, building an effective and responsive civil society presupposes certain capacities, methodology, and values that have to be adopted to achieve set goals. Table 1 identifies some of the capacities, methodologies, values that civil society ought to adopt to be effective. There are also emerging roles and contributions of civil society that were highlighted in the context of the traders' strike.

Implications for policy

The key implication for policy in the context of the city traders' strike is related to governance and leadership as key areas currently impacting on citizen's welfare in Uganda, implying the need for the Ugandan government to put in place systems or mechanisms for addressing citizens' needs and concerns. Particular attention should be paid to the social accountability aspect with regard to government interventions targeting citizens, as well as all the affected and responsible partners. The sustainability of such interventions implies process-oriented strategies and mechanisms to allow corrections and learning during the implementation of such strategies to ascertain their relevancy and significance. This has far-reaching implications for the government as the main stakeholder and partaker in development of institutions that support the functionality and engagement towards a responsive citizenry and quality service delivery.

Table 1: Emerging civil society roles and capacities

Capacities	• Platforms for the exchange of experiences and trust building towards stronger linkages in the shared purpose in civil society • A civil society that is always reflecting on the reality • Politically aware and proactive civil society • Transformative qualities in leaders at all levels • Holding others accountable • Appropriate methodology for empowerment and transformation • Building and sustaining collaborative approaches
Methodology	• A civil society that empowers community leadership • Periodic review of the "cause" and "causelessness" • Constructing spaces for civil society to learn together • Constantly affirming intentions, agency, and purpose • Civil society to create platforms for learning from outstanding leaders, such as Nyerere, Mandela, Gandhi, Obama etc, (to anchor our leadership development in civil society and governments) • Design tools for measuring total qualitative and quantitative information • Focus on capacity development to become learning and process-based • Build a critical mass of courage and commitment among various levels of civil society actors – *action* and *sustainability* • Take the risk to be vulnerable and accountable whilst holding others accountable at same time • Create spaces for civil society to influence for positive response • Asserting ourselves before manipulators when they are diverting civil society from its original intentions
Values	• Accountability • A civil society that respects, appreciates and seeks to understand and work with the local realities, experiences, and knowledge
Roles	• Listening, interpreting, questioning, educating, and strengthening • Monitoring impact and chance • Documentation • Mobilising • Influencing • There is need for civil society to be clear on what they stand for, as well as become role models to influence the choices of people towards societal change • There is need for empowerment programmes by the civil society towards enabling and facilitating independence of the target groups • Become more conscious of what civil society wants to achieve, and articulate it clearly • Articulate the citizen's needs and concerns clearly • Be pro-active, use liberating and participatory approaches
Contribution	• Hosting and facilitating open dialogues • Consciousness rising about the existence of issues affecting civil society and citizens and the dynamics • Revolutionise understanding and facilitation of participation. • Emotional intelligence content in capacity development, process interventions • Seek to influence values, and language including the naming of projects between government, civil society, and private sector
Identity	• A civil society that is characterised by transformative leadership

Notes

1. Uganda's Anti-Homosexuality Bill (often called the "Kill the Gays bill" in the media) is a legislative proposal that would broaden the criminalisation of same-sex relations by dividing homosexual behaviour into two categories; aggravated homosexuality-where an offender would receive death penalty, the offense of homosexuality in where an offender would receive life imprisonment.
2. In the post-independence period, the Obote-led UPC Government sought to control the media and to bring it exclusively under state control. The country's only radio and television stations were state owned. The newspapers that had mushroomed in the colonial period disappeared, except for a few. The trend initiated during the first Obote regime continued in the Amin period.
3. In the early post-independence period numerous strikes were organised by the working class. Invariably these were crushed by the state, which in turn enacted repressive laws in response. For instance, the Trade Union Disputes (Arbitration and Settlement) Act and Regulations restricted disputes/strikes in the essential services.
4. For instance, in 1973, the military government of Idi Amin banned all women's organisations in the country. In 1978, the National Council of Women was created under Decree No. 3 as the main national machinery for the integration of women in development and umbrella body for co-coordinating all women's organisations in Uganda. The Decree made it mandatory for all women NGOs to register with the Council. As a result of this law, most national organisations disappeared, kept a low profile or became dormant in this period.
5. The Civil Society Fair 2012 was organised by the Uganda National NGO Forum from 31 May–1 June at Hotel Africana, Kampala, to showcase civil society innovations and development approaches for appreciation by all stakeholders and to consolidate stakeholder linkages, synergy, and networking for greater impact and cohesion.
6. The walk-to-work protest, as it was called, began on 11 April 2011. A group calling itself Activists for Change (A4C) organised the demonstration following rising prices for food and fuel, while the government planned to buy eight fighter jets worth $740m. On 27 July 2011 Ugandan teachers under their umbrella organisation the Uganda National Teachers Union (UNATU) laid down their tools, demanding that the government increase their pay by about 50 per cent. Teachers, estimated to number 160,000, are some of the lowest paid civil servants in Uganda. On average, a primary teacher earns around US $100 per month, while their secondary school counterparts take home around $150.
7. The Uganda Country Reflection meeting on KACITA Strike Action was organised and facilitated by Jackline Kabahinda under the APODEA Uganda Chapter platform in February, 2012.

References

Bazaara, N. 2000. "Contemporary Civil Society and the Democratization Process in Uganda: A Preliminary Exploration." Working Paper 54." Kampala, Uganda: Centre for Basic Research.

Bazaara, N., and J. Barya. 1999. "Civil Society and Governance in Uganda: A Historical Perspective". Paper presented at the Second International Conference on Civil Society and Governance, February17–22, in Cape Town, South Africa.

Kruip, G., and H. Reifeld, edited by 2007. *Church and Civil Society: The Role of Christian Churches in the Emerging Countries of Argentina, Mexico, Nigeria and South Africa*. Sankt Augustin/Berlin: Konrad Adenauer Foundation.

Mamdani, M. 1983. *Imperialism and Fascism in Uganda*. Nairobi: Heinemann Educational Books.

Oloka-Onyango, J. 2000a. "Civil Society, Democratization and Foreign Donors in Uganda: A Conceptual and Literature Review." Working Paper 56." Kampala, Uganda: CBR.

Oloka-Onyango, J. 2000b. "On the Barricades: Civil Society and the Role of Human Rights and Women's Organisations in the Formulation of the Bill of Rights of Uganda's 1995 Constitution." Working Paper 60." Kampala, Uganda: CBR.

Tamale, S. 1999. *When the Hens Begin to Crow: Gender and Parliamentary Politics in Uganda*. Kampala, Uganda: Fountain.

URN. 2012. “Report Warns of Shrinking Civil Society Space in Uganda”. Accessed June 20, 2012. http://ugandaradionetwork.com/a/story.php?s=44765.

The author

Jackline Kabahinda is a development practitioner, and a promoter of social justice through social activism and transformational leadership. At the time of writing this article, Jackline was coordinating Civil Society Institutional Development at EASUN, The Centre for Organisational Learning based in Arusha, Tanzania, East Africa. Currently, Jackline is pursuing a PhD in Political Science at Technical University Dortmund, Germany. In her different roles she has contributed to, and championed several advocacy engagements including pioneering the Association for Practicing Organisational Development in East Africa (APODEA-Uganda Chapter) towards facilitating value based development practices. Jackline has previously worked with the Uganda National NGO Forum and Uganda Debt Network (UDN), the National Organic Agricultural Movement of Uganda where she managed advocacy programmes. She has written on corruption and land policy in Uganda and is passionate about land rights for women, an area is currently researching about in the context of the international rights frameworks and applicability in the rural context.

Civil society at multiple crossroads in Asia

Kaustuv Kanti Bandyopadhyay

Over the past two decades the centre of gravity of the global economy and politics has shifted to Asia. The region, with its demographic, economic and political vibrancy, has changed the rules of the game globally. The region also represents extreme heterogeneity and perplexing paradoxes compared to other regions of the world; it is faced with multiple crossroads, including around poverty, illiteracy, access to water, and sanitation. It the midst of this, Asian civil society is on the rise. This viewpoint outlines its notable contributions, the multiple crossroads it faces, and future avenues.

La société civile face à de multiples carrefours en Asie
Au cours des deux dernières décennies, le centre de gravité de l'économie et de la politique mondiales a évolué vers l'Asie. Cette région, avec son dynamisme démographique, économique et politique, a modifié les règles du jeu à l'échelle mondiale. Elle présente par ailleurs une hétérogénéité extrême et des paradoxes mystérieux par rapport à d'autres régions du monde ; elle se trouve à des carrefours multiples, y compris en ce qui concerne la pauvreté, l'analphabétisme, l'accès à l'eau et l'assainissement. Au milieu de tout cela, la société civile asiatique est en plein essor. Ce Point de vue présente ses contributions notables, les multiples carrefours auxquels elle fait face et les avenues futures.

La sociedad civil de Asia ante varias encrucijadas
Durante las últimas dos décadas, el centro de gravedad de la economía y de la política a nivel mundial se ha trasladado a Asia. Con su vitalidad demográfica, económica y política, la región ha cambiado las reglas del juego en todo el mundo. Asimismo, encierra heterogeneidad y paradojas extremas en comparación con otras regiones del mundo: enfrenta varias encrucijadas en torno a la pobreza, al analfabetismo, al acceso al agua y al saneamiento. En este contexto, la sociedad civil asiática sigue prosperando. El presente artículo realiza un bosquejo de sus notables contribuciones, de las múltiples encrucijadas que enfrenta y de sus caminos a futuro.

Sociedade civil em meio a múltiplas encruzilhadas na Ásia
Nas duas últimas décadas, o centro de gravidade da economia e política global tem mudado para a Ásia. A região, com seu dinamismo demográfico, econômico e político, tem alterado as regras do jogo globalmente. A região também representa uma heterogeneidade extrema e paradoxos desconcertantes se compararmos com outras regiões do mundo. Ela se encontra em meio a múltiplas encruzilhadas, inclusive a respeito da pobreza, analfabetismo, acesso a água e saneamento. Em meio a isto tudo, a sociedade civil asiática cresce. Este ponto de

vista apresenta suas notáveis contribuições, as múltiplas encruzilhadas em que ela se encontra e caminhos futuros.

Introduction

The combined forces of liberalisation, privatisation, and globalisation have fuelled the emergence of a vibrant Asian economy. However, the region still faces considerable challenges of poverty and exclusion. In the face of multiple paradoxes, civil society, once thriving and vibrant, is at multiple crossroads. The paper first illustrates various progresses made in the region; then identifies emerging developmental challenges; moves on to focus on the composition and contributions of civil societies, and some significant civil society crossroads; and finally provides some suggestive strategies for civil society to deal with these crossroads.

Asia on the rise

Over the past two decades the centre of gravity of global economy and politics has shifted to Asia. The region, with its demographic, economic, and political vibrancy, has changed the rule of game globally. The region also represents extreme heterogeneity and perplexing paradoxes.

Asia is the "new axis" driving global economic growth. It accounts for almost 37 per cent of world GDP at purchasing power parity. The proportion of world economic output generated by Asia is expected to be about 50 per cent by 2050, with China probably overtaking the US and Europe (Lee and Hong 2010). Profound economic transformations and massive migrations from rural areas have resulted in rapid urbanisation of Asian populations. Since 1990 the urban population has grown from 1 billion to 1.76 billion, and by 2030 it is projected to be 2.6 billion. Twelve out of the world's 21 megacities are found in Asia (UNHABITAT 2012).

Many Asian countries have graduated away from Official Development Assistance (ODA). As well as Asian OECD members (Japan, Australia, New Zealand, and South Korea), countries like China, India, Malaysia, Singapore, Indonesia, and Taiwan now extend development assistance to other developing countries. New formations of global governance are beginning to reshape the global political landscape. BRICS (Brazil, Russia, India, China, and South Africa), with two Asian countries – India and China – is seen as a powerful grouping of emerging economies with the imminent possibility of setting up a new development bank with global outreach. ASEAN (Association of Southeast Asian Nations) has become the most vocal subregional grouping from Asia to have a global voice. Nearly one-third of G-20 members are from Asia, which has begun to reshape global governance institutions including the International Monetary Fund (IMF) and the World Bank.

Many countries in Asia have a long tradition of democratic governance. Australia, New Zealand, South Korea, Japan, Malaysia, India, Sri Lanka, the Philippines, and Thailand have been practising democracy for several decades. Many other Asian countries like Myanmar, Afghanistan, Pakistan, Bangladesh, Nepal, Cambodia, Indonesia, East Timor, and Bhutan have been making transitions towards democratic governance. In the past decades, institutions of local governments have gained considerable mandates and resources to serve the needs of the poor and the marginalised. This trajectory of democratisation of governance presents a new opportunity for promoting greater citizen participation and civil society voices. In many Asian countries, independent media has become far more active in supporting demands for accountability. New information technology is making the communication of struggles, protests, demands, and voices far more visible and audible.

Asia at the crossroads

Asia is also home to approximately 900 million poor people, accounting for approximately half of the world's poor. The region has one of the highest illiteracy rates in the world. In Asia about 685 million people do not have access to improved water supply and 2 billion do not have access to improved sanitation. Over half of the world's undernourished people and slum dwellers live in Asia, and 43 out of every 100 children who die before reaching age five are born in Asia. Hardly any of the MDG targets will be achieved in South Asia. In South East Asia, Cambodia, Lao PDR, and Myanmar are lagging behind targets; with a similar trend being noted for most Pacific islands (UNDP 2010). The status of women, and consequently various related MDGs, continues to be appalling. More than 40 per cent of urban population lives in substandard housing (Roberts and Kanaley 2006). The Asian region is highly vulnerable to impacts of climate change and its poor households are more so due to their dependence on natural resources for survival (World Bank 2010).

It is ironic that despite impressive economic growth many Asians still face poverty, vulnerability, and exclusion. This brings attention to failure of governance and massive corruption in these societies. Most administrative structures are inefficient to deliver their mandated tasks. Decision-making about resource allocations and expenditures is opaque and archaic. Many Asian countries share a history of suppression and authoritarian regimes. Despite recent trends in democratisation and decentralisation, development programmes remain highly centralised in most countries. Reform of governance to bring about transparency and to ensure accountability of institutions and decision-makers remains a challenge in Asia today.

Many countries in the region are witnessing widespread violence. Some of it is linked to the so-called war on terror, while others have recently faced violence arising out of destruction of traditional livelihoods, cultures, and societies as the modern growth paradigm is increasing inequalities within these countries. Many Asian countries are also witnessing increased strife and violence arising from ethnic, caste, and religious conflicts. These violent conflicts have heightened the sense of insecurity among Asians.

Asian civil society on the rise

Civil society in Asia represents a wide range of formations and a rich diversity in roles. The mosaic includes community-based organisations, mass organisations, religious organisations, voluntary development organisations, social movements, corporate philanthropy, consumer groups, cultural associations, professional associations, economic associations, media, and academia (Tandon 2002). Civil society in Asia countries has contributed locally, nationally, and trans nationally in myriad forms. The following sections capture notable contributions of civil society in Asia.

Providing services to the excluded and unreached

Despite impressive economic prosperity, poverty remains pervasive in Asian societies. A considerable number of poor do not have access to basic services. As many states in Asia withdrew their commitments to provide these services, the poor remained excluded and unreached. Many civil society organisations (CSOs) stepped into the gap created to provide these basic services to poor and marginalised people.

Promoting and scaling up innovations in participatory development practices

CSOs, over the years, have been innovative in developing many development practices, keeping people's participation at the heart of those innovations. These innovations have been fostered in education, health care, agriculture, watershed, minor irrigation, and forestry, to name a few. Many such innovations have been advocated by CSOs and eventually become mainstreamed in government-supported, large-scale development projects and schemes.

Promoting economic and political empowerment of the marginalised through mobilisation and awareness building

One notable contribution of CSOs in Asia, particularly in the post-economic liberalisation period, is promoting the economic inclusion of the poor through the formation of self-help groups, micro-enterprises, and bridging communities and market. The formation and nurturing of a large number of women's microcredit groups across Asia has contributed not only to addressing economic poverty but also to redefining gender relations within the family and society. The contribution of civil society in nurturing community based leadership and organisations, along with massive awareness raising, has resulted in more inclusive politics in many Asian societies.

Promoting active citizenship and access to rights and entitlements

Civil society in Asia has catalysed the changing relationships between citizens and states. The use of new technology like social media has given a new fillip to traditional ways of mobilising and organising citizens' voices. Citizens equipped with new awareness have become more critical about any failure or unresponsiveness on the part of their governments and political class. They do not hesitate to take to the street to raise issues along with previously unknown fellow citizens, as exemplified in India, Indonesia, Cambodia, and in other countries. A new trend is the growing participation of the middle class and youth, particularly in urban areas, in a wide variety of issues including corruption, safety, dignity, and mobility.

Supporting governance reforms and democracy building with a bottom-up approach

Civil society in the region is engaged with governance institutions and processes to generate innovative solutions that work for the poor. Civil society is at the forefront to nurture demands for better governance and deepening democracy. The practice of social accountability and citizens' vigilance in democratic governance, including elections, policy making, resource allocations, expenditures, and performances, is growing and has contributed to improve governance and bottom-up democracy building.

Promoting policy discourses on public good

A significant contribution of civil society in Asia is identifying, amplifying, and advocating contemporary issues which have policy implications from the public good point of view. A variety of strategies including evidence building, policy dialogues, protests, and movements have been used by civil society to keep these issues alive in public discourse. Irrespective of varied responses from governments, which ranged from cooperative support to coercive suppression, civil society remained indomitable until the time when those issues were addressed by the governance regimes.

Asian civil society at crossroads

The ostensive success and achievement of Asian societies during past decades is not without its share of problems and issues. Economic prosperity, growing democratic aspirations, and increased global influence, along with the stark contradictions of pervasive poverty, exclusion, disparity, violence, and intolerance, have made Asian societies and polities a complex reality. Civil society within this reality is faced with unprecedented challenges. These challenges are becoming even deeper as the very success in one aspect has been aggravating the other aspects of the society; for example, the economic prosperity fuelled by free market economy and privatisation has also contributed to growing economic disparity, environmental degradation, and dislodgement of indigenous people from their traditional habitats. Creation of enormous private wealth at the cost of other public goods has created unprecedented paradoxes. Civil society finds itself at many crossroads. The following section highlights some of these crossroads that civil society faces today in many Asian societies.

Recognition and repression – two sides of the same coin

Civil societies in many countries are well recognised by national and sub-national governments for their contributions. Governments invite them not only to serve on various policy advisory forums but also to join government. In India, the National Advisory Council (NAC) was constituted with significant representations from civil society expertise. NAC has been instrumental in conceptualising, advocating, and drafting many progressive policies with positive impacts on the poor and marginalised. Many civil society leaders have joined the Philippines government. In Cambodia, government has invited civil society to contribute to designing the Sub National Democratic Development reform programmes. Due to CSO advocacy in Nepal, special rights for women, indigenous people, persons with disability, and other excluded groups were incorporated in the Interim Constitution of Nepal. A national policy on the voluntary sector in India, or the constitutional recognition of NGOs in the Philippines, represents exceptional innovation for an enabling environment for civil society.

Contrary to such positive engagements, in many Asian countries civil societies are also subjected to stifling laws and regulations, harassment, and repression, irrespective of the depth and maturity of democratic regimes. Several countries have no legal instruments for incorporating non-profit organisations. Even where laws exist, taxation laws for non-profits are ambiguous or restrictive. In India, not only were more than 4,000 Foreign Contribution Regulation Act (FCRA) registrations of CSOs cancelled in 2012 (Sebastian 2012), but the entire sector is being harassed by a new Income Tax law. In Indonesia and Cambodia, CSOs are struggling to resist the implementation of draconian NGO laws which are being considered by the governments. In many parts of Asia like India, Pakistan, Sri Lanka, Indonesia, Philippines, a large number of human rights groups are being harassed under the pretext of terrorism, secessionist movements, or political violence by the ultra-leftist groups.

Civil society-government collaboration – performing on behalf of or monitoring performance

Governments and civil societies in many Asian countries are partnering to implement government programmes and schemes and to deliver basic services. These partnerships remain acceptable as long as civil societies agree to abide by government defined rules of engagement. Most governments accept the roles of civil society in service provision, but involvement in advocacy, campaigns, and governance reforms are resented or even actively negated. Nonetheless, this has resulted in co-option of civil societies in government service delivery mechanisms; the subsequent loss of autonomy raises critical issues related to these programmes and schemes. It has also raised questions about the ability of civil society to hold governments accountable and to play watchdog functions.

Some governments are occasionally inviting civil society to monitor specific policies or programmes by gathering authentic feedback from the field. A section of government officials are amenable to third party civil society monitoring. However, these examples are few and far between. In most instances governments are apprehensive about monitoring and watchdog functions being played by civil society, particularly where civil societies have been critical about government performances. An emerging crossroads within civil society in Asia thus includes the predominant role of civil society as partners of the government to deliver services or to hold governments accountable to citizens for their conduct, performances, and commitments.

The partnership between civil society and local governments has evolved prominently in recent decades. Since early 1990s many Asian countries have been experimenting with democratic decentralisation, resulting in increased representation from hitherto marginalised groups like women, indigenous people, and other minorities. Civil societies have enhanced the capacities of these leaders and institutions to make local governance effective. However, civil society is also confronted with the dilemma as to whether these institutions are indeed able to rise above the embedded structural inequities in many Asia societies. Institutional inefficiency, corruption, nepotism, and divisive politics remain pervasive in many local governance institutions. This brings civil society to a crossroads as to whether to continue providing capacity development support to these institutions, or to the citizens' organisations to hold the former accountable to citizens.

Civil society-political society relationships: hostility or engagement

The relationships between civil society and political society in many Asian countries have been historically dynamic. Civil society participation in movements against imperialist or dictatorial regimes has always been welcomed by political leadership. Immediately following independence in India, Pakistan, and Bangladesh, and in post-revolution Philippines and Indonesia, civil society and political society reinforced each other. However, in recent decades as civil society continues to participate in movements on contemporary issues like people's rights over natural resources, or civil, political, and development rights, political society which supports the governance regime finds civil society on the wrong side. The relationship has been further aggravated with the deteriorating political culture, characterised by corruption, power seeking attitudes, and unaccountable behaviour of many political leaderships. As civil society raises concerns or mobilises citizens against such political classes, the hostility increases exponentially. Civil society in the region finds itself in a major crossroads as it needs to engage with political society in order to influence policies, programmes, and institutions for the benefit of the poor. The challenge is to remain faithful to representative democracy while advocating for participatory and inclusive democracy and politics.

Diminishing resources – upheaval task: putting the cart before the horse

A valuable benefit of civil society has always been advocacy and promotion of equity and justice within the framework of a democratic society, counteracting other power holders. Asian societies are facing inequality and injustice alongside economic prosperity for a few. Citizens aspire for more democratic order and space, and addressing these issues requires a resourceful civil society. In recent decades, as international development assistance is rapidly disappearing, a major section of Asian civil society is facing tremendous hurdles to assuming roles around advocacy, mobilisation, awareness generation, and capacity development. A major crossroad for civil society is to find new resources from the government and private corporations to nurture its relevance in society by taking sides of the poor and marginalised, whilst the very policies, approaches, and behaviours of these institutions are perpetuating inequalities and injustices.

Solidarity with humanity – domestic or international

As many governments and private corporations from Asia have become global, their influence on both developing and developed countries in other regions is also increasing. The volume of government aid from many Asian governments within and outside the region has grown many folds. India and China are becoming dominant, alongside other traditional donor countries like Japan, South Korea, and Singapore. While many such aid grants and technical assistance are well intended, many investments are also creating new forms of exploitation and inequities in other societies. Civil societies in most Asian countries are overwhelmed with domestic issues related to their own societies, polities, and governance. Another crossroad for Asian civil society therefore is to find a balance between continuing interventions in their own societies and extending solidarities to people in other societies, which may occasionally require taking a stand against their own governments.

Towards a future agenda

Protecting and advancing democratic space for civil society

A prerequisite for vibrant civil society is the availability of democratic space; this space is restricted in many countries and shrinking in others. One of the first and foremost tasks is to protect and advance the democratic space for independent, diverse, and inclusive civil society in each Asian country. Developing national policy for possible roles and contributions of civil society is one such approach; another is to modernise and liberalise regulatory mechanisms to allow for the incorporation of civil society as non-profit actors. One building block is upholding fundamental rights for citizens – freedom of association and freedom of speech – which are either not enshrined in some countries or not adequately practised in some others. Defenders of human rights and independent civil society voices need enabling and supportive policy environments. Practical steps need to be devised in each country to create and strengthen such policy frameworks and enabling legislations.

Strengthening new and old coalitions

Various innovative civil society actions in the region were characterised by working in coalition. These coalitions cut across citizens' associations, community organisations, professional NGOs, academia, media, and other groups. In some examples, coalitions included parliamentarians, local governments, and the private sector. A multi-stakeholder coalition oriented approach is not easy to develop. Even the successful existing coalitions face challenges to maintain focus and continuity. Therefore, capacity enhancement for working in coalitions and utilising multi-stakeholder approach could be planned in different countries and sub-regions.

Mobilising new resources for civil society

As many countries in the region have graduated from ODA, international funding for civil society is facing threats to its survival. The Philippines, Indonesia, and India are examples where such a threat is imminent. Government funding for civil society is available for service delivery, but not for advocacy and campaigns. Private philanthropies and corporate donors in the region prefer to support 'safe' charities rather than active and vocal campaigns. Microfinance and other business models of social entrepreneurship to generate sustainable funding for civil society operations limit engagement with governance institutions from the vantage point of the poor. There is an urgent need to generate resources for civil society actions in a creative and sustainable manner. This may necessitate reforming taxation laws to encourage greater philanthropic flows within the region and to make the growing middle class and their diasporas aware of the crucial contributions of civil society.

Engaging with new and old inter-governmental arrangements

Alongside many old inter-governmental collaborative arrangements, many new formations are emerging, either in the region or where governments from the region are joining in other global formations. Civil society needs to find ways to engage with G-20, BRICS, ASEAN, SAARC, and similar entities as the decisions made in these inter-governmental arrangements are increasingly affecting the lives of citizens. Civil society and governments need to work together to discover issues, instruments, and modalities of engagement.

References

Lee, J.-W., and K. Hong. 2010. *Economic Growth in Asia: Determinants and Prospects.* ADB Economics Working Paper Series No. 220, Manila: Asian Development Bank. Accessed January 5, 2013. www.adb.org/sites/default/files/pub/2010/economics-wp220.pdf.

Roberts, B., and T. Kanaley, eds. 2006. *Urbanisation and Sustainability in Asia: Case Studies of Good Practices.* Manila: Asian Development BankAccessed January 5, 2013. www.adb.org/sites/default/files/pub/2006/urbanization-sustainability.pdf.

Sebastian, B. 2012. "Ministry of Home Affairs Cancelled FCRA Registration of More than 4000 Organizations*e-Vani* 1 (November-December). Accessed December 1. http://vaniindia.org/pdf/November2012.pdf.

Tandon, R. 2002. *Voluntary Action, Civil Society and the State.* New Delhi: Mosaic Books.

UNDP. 2010. *Achieving the Millennium Development Goals in Era of Global Uncertainty: Asia Pacific Regional Report 2009/10.* Accessed January 5, 2013. http://content.undp.org/go/cms-service/stream/asset/?asset_id=2269033.

UNHABITAT. 2012. *Sustainable Urbanization in Asia: A Sourcebook for Local Governments.* Bangkok: UNHABITAT. Accessed January 5, 2013. http://www.unhabitat.org/pmss/listItemDetails.aspx?publicationID=3345.

World Bank. 2010. *World Development Report: Development and Climate Change.* Washington, DC: The World Bank.

The author

Kaustuv Kanti Bandyopadhyay is the Director of PRIA. He is a researcher and practitioner of capacity building, civil society, democratic governance, social accountability, participatory planning, and monitoring and evaluation. For the last 20 years, he has been associated with universities, academic institutions, and CSOs. He has a PhD in anthropology.

Civil society in changing India: emerging roles, relationships, and strategies

Debika Goswami and Rajesh Tandon

The changing socio-economic and political tableau in India has affected civil society. This paper investigates how the roles, relationships, and strategies of civil society organisations (CSOs) are changing in response to these socio-political and economic changes. It also looks at what new capacities, opportunities, and challenges are emerging as important for CSOs in these changing contexts.

La société civile dans une Inde en mutation : rôles, relations et stratégies émergents
Le tableau socioéconomique et politique en mutation en Inde a eu une incidence sur la société civile. Cet article examine la manière dont les rôles, les relations et les stratégies des organisations de la société civile (OSC) changent actuellement en réponse à ces changements sociopolitiques et économiques. Il examine également ceux des capacités, opportunités et défis nouveaux se dégageant actuellement qui sont importants pour les OSC dans ces contextes en mutation.

La sociedad civil en una India cambiante: roles, relaciones y estrategias emergentes
El cambiante panorama socioeconómico y político de India ha afectado a la sociedad civil. El presente artículo examina cómo los roles, las relaciones y las estrategias de las organizaciones de la sociedad civil (OSC) están transformándose para responder a los cambios sociopolíticos y económicos. Asimismo, analiza las capacidades, las oportunidades y el surgimiento de retos nuevos que tienen importancia para las OSC en estos contextos cambiantes.

Sociedade civil na Índia em mutação: papéis, relações e estratégias emergentes
A situação sócio-econômica e política em mutação na Índia tem afetado a sociedade civil. Este artigo investiga como os papéis, relações e estratégias das Organizações da Sociedade Civil (OSC) estão mudando em resposta a estas mudanças sócio-políticas e econômicas. Ele também examina quais novas capacidades, oportunidades e desafios importantes estão surgindo para as OSCs nestes contextos variáveis.

Introduction

Over the past two decades, Indian society, economy, and polity have all been undergoing remarkable changes, both progressive and regressive. Rapid economic growth in the past two decades following the summer of economic liberalisation (1991) has made India one of the fastest growing economies in the world. As the world's largest democracy, India has also witnessed notable achievements in the sphere of information technology. Changes in the devolution of political governance and the introduction of a series of laws like the Right to Information Act, which aim to strengthen the roots of democracy, have led to increased scope for citizen-led transparency and accountability in governance systems. Alongside developments occurring inside the country, India's influence on global governance institutions is also expanding simultaneously. However, a large section of the Indian population still faces poverty, malnutrition, and social exclusion, mainly because of an uneven distribution of economic growth. These socio-economic and political changes in India have greatly affected the country's civil society. The composition, roles, relationships, and even the resource bases of civil society are undergoing vivid changes under such changing societal circumstances. This paper investigates how those changes are likely to impact the roles, relationships, and strategies of the civil society organisations (CSOs) in the country. It also looks at what new capacities, opportunities, and challenges are emerging as important for CSOs in these changing contexts.

Changes in Indian society

In the decades immediately after independence, from the 1950s to 1980s, the Indian economy grew at an approximate rate of only 3.5 per cent per year. The crisis peaked in early 1991, following which the Indian economy was liberalised, leading to high rates of growth and rapid rise in gross domestic product (GDP). The economy is expected to double in the current decade. India will be the world's third largest economy (in terms of purchasing power parity) after China and the US by 2025 (The Island 2011). India is also the world's largest democracy with a population of 1.2 billion, half of whom are under 25 years old, a workforce often referred by demographers as the potential "demographic dividend". Rapid urbanisation coupled with remarkable advancement in the sphere of information technology (the spread of mobile telephony, satellite technology, Internet and software innovations) have also contributed to a new society.

Parallel to developments in the economic and technological domains, there are deeper changes in the political governance of the country. The introduction of democratic decentralisation and constitutionally mandated local governance systems in 1993, the enactment of Right to Information Act in 2005, the Mahatma Gandhi National Rural Employment Guarantee Act in 2005, the Forest Rights Act in 2006, and the Right to Education Act in 2009 have advanced the rights-based approach to development and the institutionalisation of many social accountability mechanisms in grassroots governance. India's influence on global governance institutions is also expanding. Its emergence as a potential super power and its expanding domestic and international market, along with its rising stature in various international forums, has established a very optimistic future. India has secured non-permanent membership on the UN Security Council and has emerged as a strong contender for permanent membership.

All these positive developments notwithstanding, a large proportion of Indians still face poverty, malnutrition, and deprivations. The benefits of economic growth have not been distributed evenly. On the contrary, the gaps between the rich and the poor have expanded. India is still stagnating in HDI ratings. This situation can make India a "demographic disaster" rather than a "demographic dividend". Urban growth in India has largely remained unplanned and exclusionary, producing problems such as the growth of slums and squatter settlements, environmental degradation, and increased burdens on urban infrastructures. Additionally, threats from extremist groups in several parts of the country have become another severe challenge for India. Pervading issues of corruption, slow implementation of governance reforms, and a sluggish bureaucratic administration have impeded the rapid reduction of poverty.

Changes in Indian civil society

The composition, roles, and relationships of civil society in India are influenced by the socio-economic changes that Indian society is witnessing. As a response, numerous forms of CSOs have undertaken a variety of interventions. It is imperative to throw some light on some of these transitions happening within the Indian civil society.

Changing composition

Over a period of two decades, since the early 1990s, there has been a "*mushrooming of voluntary organisations*" (Tandon 2002) in India. Many organisations have emerged abruptly without being aware of the local context and its needs. Many more, instead of pursuing social commitments, pursue business and commercial motivations. Another prevailing trend is for political leaders to form CSOs. Many CSOs are set up by ex-bureaucrats, ex-corporate employees, and industrialists who have no vision for development or welfare and who regard the sector from a business or profit-making perspective. Some unemployed youth view CSOs and NGOs as self-employment ventures and money-making machines. For example, after the Tsunami in 2004 enormous amount of funds gushed into Tamil Nadu for relief, so many people opened NGOs with the purpose of acquiring these funds. Consequently, a considerable increase in the number of CSOs in Tamil Nadu was reported in the post-Tsunami period.[1]

Religious foundations are also coming up with their own networks, associations, and NGOs to build public opinion behind their own interests. Overall, civil society in India today has experienced a perceptible change; earlier it approached issues from people's perspective and the state was seen as duty bound. The approach was to support the interests of the people to claim their rights in relation to the state. Now a significant section of civil society has been co-opted by the state; they act like extended hands of the state.[2]

Simultaneously, the central and the state governments in India are forming societies (a form of non-profit organisation) to implement their programmes, and so adding to the composition of civil society. Examples include organisations like National AIDS Control Organisation (NACO – a division of India's Ministry of Health and Family Welfare) formed in 1992. It is the nodal organisation for formulation of policy and implementation of programmes for prevention and control of HIV / AIDS in India.

There are now also a large number of registered groups of micro-entrepreneurs in India, popularly known as self-help groups (SHGs). The model is one of voluntarily coming together to save small amounts regularly, mutually agreeing to contribute to the common fund, and to help meet their emergency needs. Thousands of SHGs have been set up across the country by NGOs and through various government rural development programmes (Sharma 2011, 63). Women's SHGs are also playing a major role in poverty reduction and women's empowerment through financial inclusion. Though the SHG-bank linkages have contributed to improved economic conditions, there are concerns about institutional sustainability and self-reliance (Reddy et al. 2007).

Social movements, which constitute another prominent face of civil society, raise their voices against anti-people policies, for better governance and for a corruption-free society. These movements sometimes become so powerful that the government cannot ignore them. Civil society in the form of non-formal, sporadic, and spontaneous protest movements (like the anti-corruption movement at a national level, the anti-POSCO movement in Orissa, the West Bengal land movements, the water privatisation movement in Karnataka, and many more) has gained momentum in the recent past. Other associations and networks like Right to Food Campaign that are deeply involved in advocacy form another section of civil society. For example, the anti-POSCO movement in Odisha has support from many such associations and networks.[3]

Another significant trend is the development of corporate social responsibility (CSR) promoted by the burgeoning private sector. It has gained greater currency since the new Companies Act in 2012, which suggests that companies are expected to earmark two per cent of their net profits towards CSR each year. Hence, corporations have started promoting CSR and private philanthropic activities and several are also involving CSOs in implementation tasks. However, these CSR activities are largely conducted by the corporates through the formation of separate foundations or under their own banner.

Community-based organisations (CBOs) are also a significant part of the civil society, both in urban as well as rural settlements covering excluded sections of the population. CBOs are predominantly voluntary and self-funded bottom-up organisations which are often more effective in addressing local needs than the larger charitable organisations. A significant number of urban CBOs – including Resident Welfare Associations (RWAs), Neighbourhood Associations, and Apartments Associations – have emerged lately to represent the interests of the citizens of a specific urban or suburban locality in Indian cities. Citizens' protests in urban centres (like after the recent Delhi gang-rape case in December 2012) have become more assertive and widespread in recent years.

These various forms of civil society in India reveal that the sector is rich in diversity and extremely heterogeneous in nature. In fact, "*it will be wrong to homogenise the sector; rather it is important to value its diversity*".[4] But, when such diverse sets of actors, values, roles, activities are brought cumulatively under the broad banner called "civil society", then the process also carries with it the baggage of a confused identity for the sector as a whole. Thus, diversity is both a source of strength and a problem for the Indian civil society.

Changing roles

In the decades prior to economic liberalisation in India, CSOs worked for the overall development of the society, focusing mainly on uplifting the downtrodden. Since the early 1990s, when the state started withdrawing from some of its responsibilities, CSOs focused more upon governance and development. Various international organisations and UN agencies started working actively with Indian CSOs by providing aid, monitoring, and overseeing development programmes in regions hit hard by socio-economic problems. According to Tandon (2002), the three most important contributions of CSOs in national development were (i) innovation, (ii) empowerment, (iii) research and advocacy. With changing times and emerging challenges, the roles of CSOs have also been diversifying and changing.

The role of CSOs in making Indian democracy alive and participatory is noteworthy. CSOs are engaged in a wide spectrum of activities which encompass issues of governance, advocacy, policy-making, and facilitating people's participation through awareness generation (Singh 2011,14). The policy advocacy role of CSOs needs special appreciation as some important Acts like the Right to Information Act or Domestic Violence Act have been passed with support from strong advocacy initiatives from civil society. Many CSOs work closely in association with the institutions of local governance, especially since the late 1990s following the 73rd and 74th Constitutional Amendment Act that empowered local government institutions. Another new trend is the partnership of CSOs with central and provincial governments in implementing various flagship government programmes. In this context, the shift of many CSOs to service delivery and contract-based work is noteworthy. Governments steer NGOs towards service provision and administer them through tender-based approaches, inviting bids. This approach leads to cut-throat competition among CSOs for particular projects. During the process, however, the component of voluntarism is lost from the CSOs.

In the present era of globalisation and strong market forces, some CSOs play an important role in integrating excluded sections of society within the global economic chain by developing their income-generating skills and providing them with required financial services and market linkages. Again, issues of civil liberty and political assertion by the marginalised sections, including women, scheduled castes and tribes, dalits, and other minority communities, have captured a lot of civil society space and created unprecedented pressures on the state to improve responsiveness and performance. Finally, civil society in the form of citizen protest movements enables voices against the neo-liberal policies and their adverse impacts on the people to be heard.

Three kinds of changes in the composition and roles of civil society can be ascertained over the past two decades. First, a much larger number of campaigns and policy advocacy efforts have been occurring in India. These focus on rights of the excluded; they also target negative consequences of rapid economic growth and rise of the free market. Second, much larger numbers of SHGs and microfinance NGOs have gained visibility during the past decade. Government's policy to support SHGs, banks' interests in micro-lending, and international funding towards financial inclusion have contributed towards this growth in the country. Third, spontaneous and collective citizen protests have grown in urban centres. A young, educated, economically mobile urban middle-class is now more frequently demanding better performance and justice from the state. When well-connected criminals, offenders, and crooks are seen to be let off lightly by the state machinery of police and judiciary, such protests have now found a target.

Some traditional roles of Indian NGOs have also continued during this period. These include provision of basic services to the poor and the marginalised. While basic education and primary health care were more common services even before, new areas of service delivery such as water, sanitation, and housing have developed. Likewise, religious institutions in India have always played charitable welfare roles throughout history. With the growth of new gurus and cult icons, the public visibility of such charities has increased. A rise in Hindu nationalism, and recent wave of Islamist terrorism have further enhanced resources and expansion of religious charities in the farthest regions of the country.

Amongst the new trends in roles of NGOs is the growth of foundations and agencies associated with the private sector. Some of these are held by families; some are products of companies. Much of the focus of such foundations is on education and health care. However, focus on post-secondary, tertiary, and vocational education has now gained ascendancy in such philanthropies (Premji Foundation being a most visible example in recent years). In health care, focus on low-cost curative services for a wide range of diseases has increased, beyond just preventive health care as in the past. Sports and related recreational activities have also begun to find greater space in private philanthropy today.

A related emerging trend is social entrepreneurship, which encourages business models that simultaneously generate economic returns while supporting social causes and issues. There is a growing recognition that entrepreneurship needs to be promoted so that the social agenda is mainstreamed in economic development approaches.

Changing relations

With the change in India's economy, its international relations, and its geopolitical interests, CSOs have witnessed significant changes in their relationship with other sectors like government, political society, market, and even among themselves.

Relationship with the government

Since the onset of liberalisation, privatisation, and globalisation processes in India, there has been increasing interaction between the government and civil society. The Indian state, operating in multiple spheres (local/municipal, provincial, national, trans-national), plays specific functions which inevitably impact on civil society: (a) the Indian state guarantees democratic rights to its citizens, thereby providing legitimacy to all civil society functions in India; (b) it is the primary actor behind policy-making in India; (c) it also promotes national development, planning, and implementation. The state thus acts both as funder and regulator of CSOs in India, which in turn remain inextricably linked to it. Since the early 1990s many CSOs have been implementing national government programmes and collaborating with local government agencies. Furthermore, the failure of the government agencies to protect the interests of the underprivileged groups has created increased awareness about the disappointment of marginalised groups and their estrangement from the process of governance. Such situations create an opportunity for CSOs to engage with the state.

The Indian government has emerged as one of the primary funding sources for CSOs that support programmes to assist poor communities; government support of CSOs (as evidenced by the rise in the number of government-supported NGOs schemes) and increased state funding to them have become widely prevalent. The significance of the government funding is increasing as foreign donor funding is withdrawn (Kilby 2011).The Indian state has also opened up space for increasing interactions with the CSOs so far as policy-making is concerned. For instance, the National Advisory Council (NAC) set up by the United Progressive Alliance Government (UPA) in 2004 provides civil society with policy and legislative inputs to government on social policy and rights of the disadvantaged groups. It also reviews the flagship programmes of the government and suggests measures to address constraints in their implementation and delivery. Landmark legislation like JNNURM[5]/MGNREGA[6] have been influenced by the NAC. In 2007 Dr Manmohan Singh's Cabinet approved the first ever National Policy on Voluntary Sector in India, which proposes creating an enabling environment for voluntary organisations by regular consultations with various national ministries. Consequently, the Finance Ministry has been organising pre-budget meetings with selected NGOs since 2010. NGOs provided substantial inputs in the preparation of 11th and 12th Five Year Plans.

But the space for civil society in India is shrinking as well. Withdrawal of international funding sources has made CSOs increasingly dependent upon government funds and projects, consequently losing autonomy and freedom to voice pro-people demands. Acquiring government funds has become time-consuming, complicated, and competitive under the tender-based bidding approach. The government policy of granting funds only through its approved list of CSOs reinforces the problem of limited access of government funds for non-approved CSOs. Under such circumstances, strategic partnership with the government is difficult so the emphasis is on contractual agreements. The government of India also regulates the voluntary sector through the recently passed Foreign Contribution Regulation Act (FCRA) and the new Direct Tax Code (DTC). Most grassroots organisations remain unaware of these regulatory changes, and so may be harassed by local administrations or even lose their registration (Singh 2011, 15). Besides, the relationship between the newly emerged grassroots movements and the government is often strained because these movements focus on citizenship rights and criticise the government for being unresponsive.

Several conclusions can be drawn from these trends in civil society interface with government. First, the constitution of panchayati raj institutions and municipalities has led to enhanced direct engagement by NGOs to support their accountable functioning. Second, many state governments, especially those responsible for rural development and primary health care, invite more NGOs to deliver programmes and services as the government has recognised the expertise that NGOs have. Third, government funding of NGOs has increased significantly, though it is primarily for service delivery and through a tender-based competitive system. A recent study also highlights enormous corruption entailed in securing such government funding by NGOs. Fourth, policy-making interactions and dialogues between civil society and government have increased at the national level; there is now greater readiness among both parties to do so. Fifth, the regulatory framework of income tax and FCRA has tightened further, thereby making advocacy and accountability efforts of civil society subject to greater harassment and control.

Relationship with political society

According to Tandon (2011, 33), both civil and political societies are essential for inclusive social transformation in India.[7] There are instances of CSOs working as think tanks and advisors to political parties; the MNREGA and the Right to Food campaigns are clear indicators of successful engagement with legislatures by CSOs to uphold human rights and enhance living standards (Kahar 2012). Yet another example is the transformation of an NGO into a political party; the Lok Satta converted itself to a political party with the idea that engagement with active politics is probably the best option to bring about the fundamental changes needed for a new political culture. The recent declaration of the formation of a political party by Team Anna reflects similar thinking. The rising politicisation of CSOs in India, mostly amongst regional CSOs, is also worth mentioning in this respect. These CSOs, under the strong influence of the local or regional political parties, are losing their non-partisan and non-political approach.[8]

Yet, the changing nature of political parties is further constraining this interaction. The rapid criminalisation, use of illegal funds, and family control of political parties has resulted in serious difficulties in enhancing the democratic culture in politics. Most political leaders at local and district levels today use politics as a source of income and protection from the law. In their view, civil society activists are a nuisance, to be excluded and intimidated. The segment of civil society which demands accountability from political leaders faces greatest threat in this regard. Yet, some new initiatives in civil society (like Association for Democratic Reforms) are focusing their attention on making political parties more democratically accountable again.

Relationship with the market

Much of the previous generation of NGO leaders and civil society activists grew up in an era where market institutions were weak; the new era has seen a dramatic rise of private sector and market-led economic development. So, for this generation of civil society activists, the biggest challenge is to come to terms with the rise of the free market and the visibility, legitimacy, and credibility of the private sector. For a new generation of CSOs, the market is the acceptable route to prosperity, and civil society can support that process. Hence, relations with the private sector, by and large, are non-existent for a vast segment of Indian civil society.

For another section, the recent revelations of huge corruption and crony capitalism have further reinforced their belief that the market is greedy and brutally focused on profits. Problems of land acquisition and displacement of tribals and others have further accentuated this belief that the private sector causes harm to the poor and the excluded in society.

Yet, there is now growing acceptance of the roles of the market in economic development, though there is contestation about its role in social sectors. New CSR initiatives in education and health sectors are leading to some partnerships between civil society and the private sector. However, CSR programmes are mostly hitched to service delivery in very restricted geographic areas, such as the catchment areas of corporate houses to which most of their workers belong. In many instances, the so-called CSR is just the external face of corporate houses that promotes their market interest and legitimises their presence and business. Corporate houses also lack the required vision to be engaged in development activities. Hence, "*before joining hands with corporate or the private sector, it is important for the CSOs not only to review the history, rules/regulations and ethics of the former, but also to analyse whether the proposed project is designed for the best interests of the community*".[9]

In search of new resources, CSOs are also approaching private foundations and new philanthropists for support for their work. However, the nature of the relationship between civil society and the private sector is still uneasy and uncertain to result in any significant flows from the latter to the former.

Relationship with media

Twenty-first century India has witnessed a boom in electronic, print, and social media, quite contrary to the earlier dominance of the print media and government-controlled radio and television. Thus, it is extremely important for CSOs to engage with media in order to showcase their work and generate public awareness. Since the April 2011 eruption on corruption, the term "civil society" is frequently used in public discourse by the media. But the media mostly portray CSO protest activities, while examples of innovative model building or new ideas by CSOs gain little media visibility. This has created a public perception that CSOs are meant only for protests and struggles, a perception that overshadows the constructive aspects of the work of CSOs. Attempts to monitor the media are emerging through the formation of media-watch organisations like The Hoot, a not-for-profit organisation run by the Media Foundation in New Delhi.

Emerging conclusions

The previous analysis suggests that civil society in India is facing several crossroads. Historically, Indian civil society has been known for innovative experiments that could be scaled-up as national government policies and programmes, and even taken to other countries by donors and UN agencies. For decades after the painful Bihar famines in the late 1960s, the Indian voluntary sector continued to evolve in diverse ways, especially focusing on conscientisation and mobilisation of the poor and the marginalised. It brought into the public sphere issues related to tribals and dalits, women and children, workers, and the disabled. Its grassroots work highlighted concerns which had been hidden from public discourse – violence against women, environmental degradation, occupational health and pollution, child and bonded labour, displacement, and urban poverty.

The new trajectory of economic growth has placed India into a new global league. Its public revenues have quadrupled, its government programmes have increased social sector spending by threefold, and new rights to education, information, forest, and employment have been guaranteed. The world is treating India as a key economic and political player.

The previous sections of this paper have attempted to describe these changes in India and its development trajectory. These changes have altered the influences on its voluntary sector – now generally called civil society. In some ways, these changes have created new dilemmas and challenges for civil society. This concluding section identifies five key "crossroads" posed by these shifts; the choices made on these crossroads will shape the trajectory of India's civil society in the next decade.

Diversifying composition, confusing identity

The overwhelming diversity in forms, purposes, sizes, and activities of actors in civil society is amazing. The rate and frequency at which citizens in all locations of society are organising themselves to both demand and construct solutions to their daily problems of life has increased dramatically. New legally incorporated entities at the grassroots level have grown enormously (there must be nearly two million SHGs in the country already). CBOs of the excluded sections of population have multiplied several fold amongst the dalit and Muslim communities. Even middle-class neighbourhoods have many more associations. According to an estimate by the Central Statistical Organisation, Government of India (March 2012), registered non-profit institutions have grown almost three times the figure last estimated at 1.2 million in 2001 by PRIA. Intermediate NGOs have grown to reach about five times the figure last estimated at 1.2 million in 2001. Corporate foundations and CSR agencies have come into being. A large number of private schools, colleges, and universities have also been established, all as non-profit organisations. Even government agencies have been setting up NGOs and institutional mechanisms for efficient service delivery.

In this morass, diversity is both a strength and problematic. The variety of experiences, formations, and actors brings richness to the task of mobilising citizen energies in today's India. However, at the same time, it also confuses the identity of the sector. Is this all one sector? Does it really have a shared vision of society? Are some common values shared by all actors of civil society? The confusion around identity is experienced within as well as perceived from outside.

Demands for service provision and citizens' protests

The huge resources made available through government programmes are still not reaching the poor and the excluded. Senior officials and many political leaders have accepted the large-scale weaknesses in the delivery of services "the last mile" to the poorest. Resources remain unutilised, not just mis-utilised. On a regular basis, national and provincial governments are seeking out established NGOs and their networks to take responsibility for service provision. Even local SHG federations have been handed such functions as midday meals for primary schools.

Yet, the larger question of reforming the government's system of service delivery requires consistent external pressures. It is in this respect that protests are occurring throughout the country. Media has expanded its outreach and mandates. Hundreds of news channels, radio stations, newspapers, and magazines are constantly reporting on these protests. The system of governance is facing the twin crises of inadequate service provision and unending citizen protests.

Interlinked as the two are, many civil society actors want to find ways to make basic services available to the hitherto unreached sections of the population. Yet, the passage to service provision is fraught with numerous risks, as short-term protests do not necessarily reform a bureaucracy in an old democracy like India. What is the choice then? Will working to make resources and services reach the excluded absorb the actors of civil society into the system itself? Will external pressure to reform take decades, during which numerous generations will continue to suffer? The choices are not simple, or easy.

The need to engage, the difficulty in doing so

The need to engage the government at all levels has never been greater in India. There is growing space and opportunity for engagement with local governance institutions (panchayats and municipalities), line departments, and ministries. Spaces and mechanisms for engagement with government in project design, planning, monitoring, and evaluation have expanded enormously. Several civil society actors are also vigorously engaging with statutory institutions like commissions, parliaments and legislatures, governors and the president. Engagements with police, tax authorities, public media, infrastructure agencies, judiciary, and public sector undertakings have grown considerably. The National Policy on Voluntary Sector further encourages widespread engagement throughout the government system.

Yet it is not easy to engage. The political space for questioning and criticising government, its officials, and political leaders is shrinking. Corruption at all levels colours any serious engagements, as illegitimate and illicit interests are exposed and denounced. In the context of economic growth, conflicts around natural resources and displacement are labelled as incitements against national interests. Freedoms of association and speech, enshrined in India's constitution, are frequently restricted (recent pronouncements by some ministers against social media are dangerous in this regard). The difficulties in ensuring meaningful and sustained engagement that results in concrete benefits to the poor and the excluded are closely related to the lack of administrative and governance reforms in the country. Even when officials show interest in working together, frequent transfers between positions derail the process regularly. Old rules and procedures continue to co-exist as new laws and methods are invented: The Official Secrecy Act and Right to Information co-exist, just as written precedent and e-governance co-exist.

More than 25 years ago, PRIA convened an international dialogue on government-NGO relationships. Its report was entitled "Government-NGO Relations: A Source of Life or A Kiss of Death?" That dilemma has further accentuated in India today. The choices for civil society are not simple – engagement is a must, but what are the risks? What capacities are needed to engage? Where not to engage when is not obvious?

Rich India, poor India: who will fund?

In the past decade, international agencies have rapidly closed their offices and operations in India. Even international NGOs have reduced their operations dramatically. India has developed. Parliaments and citizens of OECD countries do not want to send any aid to India since its private sector is buying up companies in their homelands, and its IT companies have captured the global market. India has been multiplying its millionaires at a faster pace than anywhere else.

The dilemma of resourcing its operations is most acutely experienced by the intermediate civil society, or development NGOs, or voluntary development organisations. For most CBOs, internal membership support is adequate. Many access government funds too. Those who are focused on providing services or delivering government schemes seem to have enough resources. The ones who generate market-linked solutions (like microfinance) can use premiums to cover staff costs and administrative overheads. For some fortunate ones, large endowments, corpus, or immovable assets generate a regular income stream to allow them to pursue their mission. Pay-by-services (like schools and clinics) have also found cross-subsidisation models to enable them to work with the poor.

However, flexible funding that allows a certain degree of autonomy from government and political system has particularly dried up for intermediate NGOs or voluntary development organisations who mobilise communities to claim their rights, and whose efforts question the absence of democratic accountability of the state and its institutions.

Increased philanthropy is not quite filling in this gap, yet. Increased middle-class donations are focused on basic services to the poor (education, health care, and water); more domestic and international (diaspora) giving is also going to religious and semi-religious parts of civil society. So, what are the choices to resource underfunded streams of work? What are the critical implications of exercising any of those choices? How will the independent space for autonomous civil society be sustained and nurtured over the coming decades?

Engaging the private sector: conflict, cooperation, merger, joint venture

One of the biggest challenges facing different segments of Indian civil society is their relationships with the private sector. For-profit private businesses are now a major part of the Indian economy, society, and polity. Business leaders have been incorporated in many committees of the government to advise, plan, and monitor public investments and programmes, including those in the social sectors. Therefore, the option of apathy towards or ignorance of the business sector by the civil society is not really a choice today.

Yet the choices for engagement vary considerably, depending on the nature of private business and the perspectives and programmes of civil society actors. In respect to land acquisition, displacement from access to natural resources and destruction of life and livelihood of the tribals and other rural people, many CSOs have been in conflict with private business. This conflict is rooted in the basic choices for policy of development, where natural resources are harnessed for rapid economic growth.

For private business in education and health services, there have been many models of cooperation, where complementarities are used to advance the joint interests of civil society and private sector. Likewise, in microfinance and livelihood activities (like agro-business), there are several examples of long-term partnerships, and even joint ventures. Here the nature of business and its profit-making activities do not come into direct conflict with the activities of civil society; instead there is potential for synergy.

The growing buzz around CSR has many variations, and civil society is to make considered choices. After the new Companies Bill, it is likely that the volume of funds available per annum may reach US$ 5 billion in 2013. There are lots of potential opportunities for synergy in CSR, but the general experiences of civil society with CSR have been somewhat disappointing. Most efforts by corporates have been to spend more resources in advertising their work in CSR than actually benefitting the community. Appreciation of CSR as an investment in a future India is rather limited, and civil society has found itself marketing the brand, products, and services of their corporate partners. Going beyond the funding aspect, partnership between civil society and private business can contribute significantly to enhancing more inclusive governance and development in the country. After all, the seventeenth-century crusade for rule of law in Europe was mounted by private capital to protect itself from idiosyncrasies of the ruling elites!

Notes

1. Mentioned by a Chennai-based CSO worker and also by a professor of Public Administration from the University of Madras.
2. Mentioned by an Andhra Pradesh-based CSO leader.
3. The US $12 billion Pohang Iron and Steel Company (POSCO) project from South Korea in the state of Odisha, India needs 4,004 acres of land, of which ten per cent belong to the cultivators. Opposing the move by the Government of Odisha to allocate lands (mostly cultivable and forest lands) to the

company, people of different formations including the locals, have joined hands under the banner "POSCO Pratirodh Sangram Samiti (PPSS)" to oppose the proposed plant in the region. For more see http://sanhati.com/articles/2170/

4. Mentioned by a Bhopal, Madhya Pradesh-based CSO leader.
5. Jawaharlal Nehru National Urban Renewal Mission is a city-modernisation scheme launched by the Government of India under the Ministry of Urban Development, aiming to create economically productive, efficient, equitable, and responsive cities.
6. The Mahatma Gandhi National Rural Employment Guarantee Act (2005) is an Indian job guarantee scheme providing a legal guarantee for 100 days of employment in every financial year to adult members of any rural household.
7. "Political society" here refers to the political leadership, political parties, and/or any other organisations following the ideology of any political establishment.
8. Mentioned in the Roundtable Discussions conducted in various states of India to develop an understanding of the state of the CSOs in India today.
9. Mentioned by a Bhopal, Madhya Pradesh-based CSO leader.

References

The Island. 2011. "Visits by Big Five Leaders in Six Months Indicate India's Growing Global Stature." *The Island,* January 7, 2011. Accessed June 4, 2012. www.island.lk/index.php?page_cat=article-details&page=article-details&code_title=15324.

Kahar, T. 2012. "Power of NGOs." South Asia Global Affairs, February 2012. Accessed June 4. www.saglobalaffairs.com/back-issues/1137-power-of-ngos.html

Kilby, P. 2011. *NGOs in India: The challenges of women's empowerment and accountability*. New York: Routledge.

Reddy, C. S., G. Bhaskar Rao, S. Ramalakshmi, V. Samatha, S. Vanaja, M. K.S. Murali Krishna, K. Raja Reddy, et al. 2007. *SHG Federations in India*. Hyderabad: APMAS.

Sharma, K. 2011. "Small Loans, Big Dreams: Women and Microcredit in a Globalizing Economy." *Economic and Political Weekly*, Vol. 46 43, 58–63.

Singh, J. M. 2011. "Future of Voluntary Sector – An Indian Perspective." *Civil Society Voices*. Mar: 14–18.

Tandon, R. 2002. *Voluntary Action, Civil Society and the State*. New Delhi: Mosaic Books.

Tandon, R. 2011. "Reforming Governance in India Today." *Yojana,* November: 33–34.

The authors

Debika Goswami is a Programme Officer at the Society for Participatory Research in Asia (PRIA), India. She holds a Master's degree in History from Presidency College, University of Calcutta, and a Master's degree on "Non-Farm Activities and Rural Development in Developing Countries" from the University of Rome, La Sapienza. Debika has experience in research and documentation on livelihood/enterprise promotion and income generation of the rural women. She has also been involved in a livelihood promotion and income generation project involving the women with disability dwelling in the slums of Kolkata, West Bengal.

Rajesh Tandon (corresponding author) is founder-President of the Society for Participatory Research in Asia (PRIA), New Delhi. Over the past three decades, under his leadership, PRIA has extensively engaged in building the capacities of civil society organisations around the world. Rajesh has an electronics degree from IIT (Kanpur), an MBA from IIM (Calcutta), and a PhD in Organisational Behaviour from CWRU (Cleveland). He provided foundational leadership to International Forum of Capacity Building of Southern NGOs (1997–2001), CIVICUS (1993–99), FIM-Forum for Democratic Global Governance (1998–2011), and Voluntary Action Network India (1988–2008). He has written extensively, with manuals, research papers, books, and training materials on civil society, capacity development, participatory research, and democratic governance.

Changing civil society in Cambodia: in search of relevance

Kaustuv Kanti Bandyopadhyay and Thida C. Khus

After decades of civil war and internal conflicts, Cambodia is on the path of democracy. The country, despite steady economic growth in the last decades, is facing challenges around poverty, exclusion, and inequality. The contributions of civil society in the past two decades, in development and deepening democracy, were significant, but civil society is now facing new challenges in the face of pro-neoliberal policies of the State. The relationships within civil society are characterised by both cooperation and competition, but its relationships with political society, media, and academia are underdeveloped. Civil society's strategies and capacities require transformations to continue to be relevant in the future.

Société civile en mutation au Cambodge : en quête de pertinence

Après plusieurs décennies de guerre civile et de conflits internes, le Cambodge est sur le chemin de la démocratie. Ce pays, malgré la croissance économique constante qu'il a affiché au cours des dernières décennies, est confronté à des défis relatifs à la pauvreté, à l'exclusion et à l'inégalité. Les contributions de la société civile au cours des deux dernières décennies, en termes de développement et d'approfondissement de la démocratie, ont été considérables, mais la société civile se heurte maintenant à de nouveaux défis face aux politiques néolibérales de l'État. Les relations au sein de la société civile se caractérisent par la coopération ainsi que la concurrence, mais ses relations avec la société politique, les médias et le milieu universitaire sont sous-développées. Les stratégies et les capacités de la société civile requièrent des transformations pour continuer à être pertinentes à l'avenir.

La cambiante sociedad civil de Camboya: en busca de la relevancia

Tras varias décadas de guerra civil y de conflictos internos, Camboya ha tomado el camino de la democracia. A pesar del sostenido crecimiento económico que se ha registrado durante las últimas décadas, el país enfrenta situaciones de pobreza, de exclusión y de desigualdad. Las contribuciones de la sociedad civil para impulsar el desarrollo y la democracia han sido significativas. Sin embargo, actualmente la sociedad civil enfrenta retos nuevos vinculados a las políticas neoliberales del gobierno. En el seno de la sociedad civil, las relaciones se caracterizan tanto por la cooperación como por la competencia. No obstante, sus relaciones con la sociedad política, con los medios y con la academia están poco desarrolladas. Las estrategias y las capacidades de la sociedad civil tendrán que transformarse para ser relevantes en el futuro.

Mudança na sociedade civil do Camboja: em busca de relevância
Após décadas de guerra civil e conflitos internos, o Camboja está no caminho da democracia. O país, apesar do crescimento econômico constante nas últimas décadas, está enfrentando desafios relativos à pobreza, exclusão e desigualdade. As contribuições da sociedade civil nas duas últimas décadas para o desenvolvimento e fortalecimento da democracia foram significativas, mas a sociedade civil está agora enfrentando novos desafios em face de políticas pró-neoliberais do Estado. As relações dentro da sociedade civil são caracterizadas por cooperação e competição, mas suas relações com a sociedade política, mídia e academia estão pouco desenvolvidas. As estratégias e competências da sociedade civil exigem transformações para que continuem a ser relevantes no futuro.

Changing contexts in Cambodia

Following decades of internal conflicts and international isolation which ended in 1991 with the signing of Paris Peace Agreement, Cambodia has travelled an impressive path of reconstruction and development. The journey from a post-conflict to a democratic society has contained many hurdles. The three-fold transition – from civil war to peace, from one-party rule to multi-party democracy, and from economic isolation to regional and global integration – has given rise to complex forms of state-society relationships (Malena and Chhim 2009).

Cambodia has been experiencing a rapid economic growth in the last two decades, with an average annual Gross Domestic Product (GDP) growth of 6–7 per cent. This rapid economic growth, however, has been narrowly based in a few sectors such as garment production, construction, agriculture, and tourism, and has had little impact on poverty reduction. The benefits of growth have been unequally distributed in selected geographic areas and people. One-third of the population is below the poverty line and 15–20 per cent is in extreme poverty (World Bank 2005).

After the first multi-party election in July 1993 the Constituent Assembly adopted the new Constitution. Subsequently, three National Assembly elections, two Senate elections, three Commune/*Sangkat* Council (Communes Councils are the local governance unit in rural areas and *Sangkat* Councils are the local governance units in urban areas) elections, and one Sub-National election (Provincial and District) have helped the electoral democracy put down roots. The government, with assistance from international development partners, has embarked upon significant governance reforms – articulated in the Rectangular Strategy in 2004 (Royal Government of Cambodia 2004, 2005) and reaffirmed in the Rectangular Strategy Phase II in 2008. The strategic framework on Decentralisation and De-concentration Reform in 2005 and the National Programme for Sub-National Democratic Development (2010–2019) have opened up opportunities for strengthening local democracy and citizen/civil society participation (Royal Government of Cambodia 2010).

Civil society organisations (CSOs) are some of the most important factors in the development and democracy building in Cambodia. Although traditional forms of civil society have always existed in Cambodian society, the modern incarnation of CSOs has largely emerged after the signing of the Paris Peace Agreement in 1991 when the United Nations Transitional Authority in Cambodia (UNTAC) started mediating the democratic transition in the country after decades of civil wars. On this account, the existence and functioning of modern CSOs is only two decades old, notwithstanding the fact that some CSOs existed before and during the civil wars in Cambodia. Over the last two decades CSOs have made significant contributions and assumed catalytic roles in rebuilding and reshaping Cambodian society, polity, and economy. Such roles historically included peace building and reconciliation, service provision, community development, human rights watchdog functions, engaging in governance reforms, and bottom-up democracy building.

The contribution from the international development community, either in the form of direct operations or providing support to the emergence of indigenous Cambodian CSOs, has been noteworthy in the last two decades. Understandably, the ensuing changes in the global and regional arena, particularly with international development aid, have influenced the functioning of CSOs in Cambodia. As the major sections of Cambodian civil society and CSOs are dependent on the international donors, the emerging priorities of the latter have also influenced the functioning of the former.

Despite the practice of limited democracy and dependence on external financial assistance, the state is gaining more confidence and becoming stronger in asserting its authority. Such assertions at times are also manifested in encroachment on democratic spaces for civil society. The current hegemonic dominance of a single party often determines the quality of political space for civil society.

The aspiration of the Cambodian government to pursue higher economic growth through private investments has led to the growing extraction of natural resources, particularly lands, forests, and water. The consequent evictions have alienated a large number of Cambodians from their roots and their livelihoods. A number of CSOs and citizen movements that have criticised the government and its policies are experiencing retaliation. It is not uncommon to hear phrases like "anti national," "agents of opposition," "destabilising forces," or "excitement" used to label such civil society movements.

In light of this background, the study focuses on the intermediary CSOs to investigate how their roles, relationships, strategies, and capacities are changing in response to the socio-political and economic changes. The study also looks at new opportunities and challenges that are emerging for CSOs.

Changing forms, compositions, and roles of civil society in Cambodia

The forms, compositions, and roles of civil society in Cambodia have been shaped by its political history, particularly the violent political struggles and civil wars in the decades between the 1960s and 1990s. These struggles had enormous implications for individuals, families, and other social ties in civil society. One of the most significant impacts was the decimation of most traditional forms of civil societies, including community-based, membership-based, and faith-based formations. During these decades international relations were dominated by military ties, leaving meagre scope for development aid. While CSOs in other Asian countries were supported by international development aid, the donor community imposed a development aid embargo on Cambodia for almost the entire 1980s. External assistance from China and other communist countries was available only to government agencies (Barton 2001). The Cold War limited the growth of externally funded CSOs. Only a few international NGOs provided relief and humanitarian aid (Mysliwiec 2004). The international aid embargo was lifted with the signing of the Paris Peace Agreement under the close supervision of UNTAC in the early 1990s. So, the growth of CSOs in Cambodia started in the early and mid 1990s.

Civil society in Cambodia is dominated by intermediary CSOs, and the number of CSOs has grown exponentially since the middle of the 1990s. These CSOs perform many functions and diverse roles. In Cambodia, international NGOs (INGOs) register with the Ministry of

Foreign Affairs (MOFA) and local CSOs register with the Ministry of Interior (MOI). A study conducted by the Cooperation Committee for Cambodia (CCC) in 2011 recorded 3,492 registered local and international organisations, of which 508 were INGOs listed with the MOFA. The report further observed that not all the registered organisations were active: an estimated one-third of the registered local CSOs and slightly more than half of the registered INGOs remained active in 2011. It also found that the CSOs spent an estimated US $550 million in the social sector, employed approximately 52,650 Cambodians, and benefitted between 2.8 to 4.5 million Cambodians directly (CCC 2012).

Since the early 1990s INGOs have occupied significant civil society space in Cambodia, but in the last decade many have facilitated "indigenisation" of Cambodian CSOs. They have supported Cambodian nationals in senior INGO project positions to form new organisations. These erstwhile INGO staff now lead many Cambodian CSOs. The Cambodian Centre for Study and Development in Agriculture (CEDAC), for example, was established in 1997 with support from a French NGO called Group for Research and Exchange of Technology (GRET). Some of the Cambodian nationals employed, including the current Executive Director of CEDAC, were GRET project staff. As GRET decided to withdraw its direct presence in Cambodia, it supported the continuation of its interventions through CEDAC. Over the years CEDAC has evolved into an autonomous CSO reaching hundreds of farmers in Cambodia. CEDAC works to improve agricultural product through new rice production technology, and through capacity development of rice farmers in Cambodia. CEDAC has organised farmers associations and supported them to produce organic rice for local consumption and export. CEDAC has successfully raised funds for the different associations to have their own rice mill, so they can cut out the middle persons and maximise the direct benefit to the farmers themselves.

Trade unions are the largest membership-based CSOs in Cambodia. Like intermediary CSOs, trade unions and labour movements are also young in the country. It was only in 1997 that the promulgation of Cambodian labour laws created legal space for the exercise of freedom of association, the right to collective bargaining, and the right to strike. Subsequently, a number of trade unions emerged, some linked to political parties and others claiming independence. The number of trade unions is increasing as internal splits produce new unions. It is estimated that only one per cent of the total workforce belongs to trade unions or associations. The vast majority of organised workers work in the garments industry, where about 60 per cent of workers are unionised. Unionisation is on the rise in the construction, transport, hotel and tourism, and rubber plantation sectors. Civil servants and informal economy workers are not allowed to organise into unions but can form associations. In 2011 the government, with the assistance of the International Labour Organisation (ILO), proposed amendments to the trade union law. However, a number of trade unions expressed concerns with the amendments, which ostensibly give undue advantage to the industries' owners, and are still under consideration by the Ministry of Labour (Nuon and Serrano 2010).

A good number of CSOs are engaged in roles such as peacebuilding, reconciliation, and mediation at the community level, although Cambodian society has moved from being a typical post-conflict society. They carry out these early roles as well as more contemporary roles such as community development or service provision. Youth for Peace (YFP) works with high school students and local communities to collect stories and testimony of what happened during the Khmer Rouge era. YFP has supported the communities in building community museums to document and display these testimonies so as to educate the next generation on the understanding of what went wrong during the Khmer Rouge time. Youth Resources Development Programme (YRDP) works with university students to strengthen communities' services and leadership, organised across countries to develop a peace curriculum for youth in the region.

Many intermediary CSOs engage in service delivery activities, such as running health clinics or schools. As government funding of public services such as health and education is meagre, these CSOs fill important gaps for underserved communities. Often they partner with local governments, particularly with the Commune/*Sangkat* Councils, to make such services available. Kampuchean Action for Primary Education (KAPE), for example, provides support to more than 300 schools, including scholarships to a large number of students from poorer families. Live With Dignity, previously known as the Lutheran World Federation, works in several provinces through an integrated development programme that includes building schools, credit programmes, increase agriculture production, and business development. LWD also organises village rice banks and promotion of women empowerment programmes in their targeted areas in Cambodia.

Another important sub-section of CSOs intervene to promote sustainable livelihoods. These CSOs provide agricultural inputs to farmers and fishing equipment to fisher folks, and so on. A related intervention is promoting and supporting organisations of farmers or fisher folks and self-help groups for of women. Microfinance activities are not as strong as in many other Asian countries and primarily used as mobilisation strategy. Such mobilisation efforts are largely limited to the local level, except for a few CSO support organisations that promote networks and coalitions of producers' organisations at the provincial and national levels. For example, Fisheries Action Coalition Team (FACT) works closely with fishing communities around Tonle Sap Lake, in Cambodia's upper and lower Mekong regions, and the coastal provinces. It supports them in building organisations and networks so that they can advocate effectively for themselves on issues that affect them. FACT has catalysed the formation of a national network of fisher-folks called Cambodian Coalition of Fishers (CCF) comprised of three regional fisheries networks – Coastal Fisheries Network (CFN), Mekong Fisheries Network (MFN), and Tonle Sap Fisheries Network (TFN). Many intermediary CSOs also facilitate the linkage of producers' groups to markets. These CSOs offer information mediation as well as material support to producers' groups. The information ranges from advanced practices on farming, animal husbandry, and fisheries, to market prices and trends. For example, Farmer Livelihood Development (FLD) promotes Village Market Networks (VMNs) of farmers and provides them with support to improve agricultural practices and marketing skills for their agricultural products.

As this support involves the use of natural resources such as land and water by the community, a few CSOs have been involved in confrontation with government agencies and powerful business interests that seek to appropriate those natural resources. In recent years the fisher folks engaged in small and medium-scale fishing are losing their livelihoods as a result of large-scale mechanised fishing. Government authorities have failed to protect the interests of these small fishing communities, and CSOs offered them organisation building and advocacy support.

Similar confrontations are emerging on the issue of land. The government is providing land concessions to domestic and foreign investors by creating special economic zones, converting huge tracts of agricultural lands for industrial and commercial activities. These projects have destroyed the livelihoods of many farmers who were already impoverished. Peoples' movements are erupting to protest against land acquisitions without proper rehabilitation and compensation. The issue of land is further complicated by the Khmer Rouge regime's prohibition of private land ownership. Although the land has been utilised by community members since the fall of the Khmer Rouge, the land titles were not properly registered, so it is difficult for illiterate farmers to seek justice through the judiciary system. Many such issues are being taken up by CSOs to provide legal assistance and mobilisation support to dispossessed farmers. For example, since 2002 Community Legal Education Centre (CLEC), has provided legal assistance to thousands of rural Cambodian families. It not only educates these families on legal rights to land and other natural resources, but also provides professional support to access judiciary systems. Its work focuses on victims of economic land concession, and domestic migrant workers.

Since 2000 some CSOs, with donor support and recognition by government policy such as the Rectangular Strategy, have worked on governance reforms, focusing on democratic decentralisation, social accountability, and reduced corruption. As Cambodia has decentralised governance through direct elections at the Commune/*Sangkat* Council level and indirect elections at the District and Provincial Council levels, many CSOs seized the opportunity to enhance citizen participation in local decision-making. These CSOs have mobilised citizens in preparation of Commune/*Sangkat* level plans as these are the basis for effective public service delivery. CSOs have made such planning processes more participatory and transparent. A number of organisations like Pact Cambodia, AMARA – Cambodian Women's Network for Development, and Ponleur Kumar provide technical support to Commune Councils in the preparation of Commune Development Plans and Commune Investment Plans. Although decentralisation is now rooted following three consecutive elections of Commune/*Sangkat* Councils, the devolution of authority related to functions, functionaries, and funds is still restricted and controlled by centralised line departments. As the newly elected members do not receive regular capacity development support from the higher tier of governments, many CSOs help Council members to clarify the roles and responsibilities of Commune/*Sangkat* Councils, organise Council meetings and open public forums, and facilitate participatory planning and budgeting. These CSOs play important roles in enhancing capacities of democratic institutions such as Commune/*Sangkat*, District, and Provincial Councils. For example, SILAKA provides capacity development training to the Commune/*Sangkat*, District and Provincial council members, particularly to women council members, and Provincial and District Governors on their roles and responsibilities, participatory planning, and budgeting, and laws related to local governance institutions.

Of late, many CSOs are engaged in social accountability practices through citizen monitoring of government development activities, particularly on public service delivery. This is a new role for CSOs that has support from the donors. These CSOs often facilitate dialogue amongst citizens, and elected and government officials on development priorities and issues. Given the hierarchy-conscious political and bureaucratic culture in Cambodia, such intermediation is a difficult but important intervention. The Demand for Good Governance project of the World Bank, for example, provides support to a number of CSOs to work on social accountability. Most of these organisations have been facilitating interface dialogues between citizens and local governance institutions on a variety of issues like improving public service delivery, access to land, water, forests, and other natural resources.

Another professional group of CSOs promote and protect human rights. These CSOs are critical about many governmental policies and programmes which encroach upon citizens' rights or curtail freedom of speech or association. A number of CSOs like The Cambodian Centre for Human Rights (CCHR), Cambodian League for the Promotion and Defence of Human Rights (LICADHO), Cambodian Human Rights and Development Association (ADHOC), and many other CSOs are prominent organisations working on human rights issues. They monitor human rights abuses and provide services to victims of human rights violations.

The current composition of CSOs and the roles played by them are shaped by both the political history of Cambodia and the international development aid community, particularly after UNTAC led its political and democratic transition. In summary, CSOs are playing the following roles:

- Providing relief, rehabilitation, and humanitarian assistance
- Provision and implementation of services
- Community mobilisation and organisation building
- Capacity development, awareness generation, and information dissemination
- Knowledge generation through action research
- Convening policy dialogue and advocacy

Changing relationships

Relationships among civil society organisations

The civil society landscape in Cambodia is dominated by intermediary CSOs which largely depend on international aid. In a sense the relationships among Cambodian CSOs and between Cambodian CSOs and INGOs are also influenced by the nature of international aid and the associated practices. At present, the relationship amongst CSOs is characterised by both cooperation and competition.

Almost all the CSOs interviewed for the study on which this paper is based belonged to networks or coalitions at the national or provincial level. These networks and coalitions provide information and capacity development support, undertake research, policy dialogue, and advocacy on issues that concern their members. The most prominent networks at the national level include NGO Forum on Cambodia, Cooperation Committee of Cambodia (CCC), ADHOC, and COMFREL. Some of them provide funding support to members; for example, MEDICAM, a network of CSOs working on health issues, provides project funding support to its members. Although funding support creates a donor-recipient relationship to smaller CSOs, that role allows bigger organisations to handle project management and donor relations.

Many INGOs promote networks of CSOs among their partners for the purposes of learning, sharing, and coordination. As some of these networks are organised around projects their composition changes as the projects change. Nonetheless, such networks have created "relationship capital" among the CSOs. PACT Cambodia, for example, during the implementation of a major project called "Local Administration and Reform Programme (LAAR)", catalysed a network of 13 Cambodian CSOs who were involved in the implementation of the project. The project aimed to enhance capacities of CSOs and Commune Councils to deepen participatory democratic practices in the local governance institutions.

CSOs operating at the local level has also catalysed provincial networks of CSOs. Typically, some prominent CSOs with greater capacities and relationships with national networks have taken the lead in creating provincial networks. AMARA, for example, in Battambang province takes the lead in organising the CSOs working in this province to form an informal provincial network. AMARA has systematised the list of provincial CSOs and often facilitates meetings and workshops to disseminate relevant information to these CSOs in Battambang. Smaller CSOs find it useful if they lack the connections to get relevant information, and if the bigger CSOs take on information bridging and capacity development roles.

Generally speaking, over the last decade there has been a proliferation of CSO networks and coalitions both at the national and provincial levels. While many CSOs justified joining networks to build solidarity and to overcome information isolation, some are very critical about the utility of such networks. They argue that the "*information and resource asymmetry*" produces power differences that work in favour of the larger organisations. Some view it as "*patronising*" and "*not helpful*", especially when smaller organisations need to confront or criticise government policies or locally powerful politicians (Field Interviews).

Historically, most intermediary CSOs are engaged in service delivery and implementation rather than the promotion and empowerment of community based organisations (CBOs). Large-scale mobilisation of citizens in the current political context is seen as threat to the political establishment. Despite the restoration of peace and establishment of democratically elected governments at national and local levels, CSO mobilisation of citizens is often seen as political activism that transcends perceived roles of civil society. Nevertheless, some CSOs have formed local CBOs to promote sustainable community development, economic empowerment, livelihood promotion, or environment protection and conservation. These CBOs receive support from CSOs for project activities, but they are also increasingly receiving "civic educational" support on topics such as rights and entitlements of citizens. In the current political culture this subtle education in citizenship and democratic rights makes sense, as it evokes less suspicion and repercussion from the political establishment. Such support, though minimal at present, has the potential for authentic community empowerment.

A number of conversations revealed that the current relationship between intermediary CSOs and CBOs is typically a patron-client relationship. As community empowerment has not been a CSO project priority, CBOs formed under projects often dissolve once the projects are over. Some CSOs are also over-possessive of their CBOs. Some CSOs become disappointed and even get angry when other CSOs tried to work with 'their' CBOs. As a result it is not uncommon to see village leaders who participate in several CBOs started by different CSOs. While the multiplicity of CBOs may enable more opportunities to multiply local leaders, they also create potential for overlap, contestation, and conflict.

While most CBOs are formed at the village or commune level, a few CSOs have also tried to create national level CBO networks. Such efforts have dramatically improved the prospect of issue-based or interest group-based mobilisation, advocacy, and policy influence at the national level.

The current political context, which is characterised by one-party hegemonic political regime, seems to be quite hostile to social and political mobilisation and movements. The current economic development paradigm of free market and foreign investments (primarily from China, South Korea, Malaysia, and Singapore) has resulted in large-scale evictions, denial of rights, and loss of livelihoods for many Cambodians who depend on traditional forms of livelihood linked to land, forest, and water. In such situations, criticism and protest against governmental policies and programmes are not uncommon. However, given a generally closed political establishment, these movements and protests are seen as undemocratic, conspiring, and destabilising forces. The CSOs are well aware of the political situation and are often empathic to social movements. In almost all interviews CSO leaders expressed in private their moral support for movements against injustices to the community, though only a few dared to provide explicit support and expressions of solidarity. The recent killing of the environmental activist Chut Wutty by military police and the detention of women activists by police became a rallying point for civil society to condemn these incidents, but only a few CSOs raise their voices in public.

Relationship with government

The relationship between CSOs and government in Cambodia has many hues and colours. It has been influenced by their expectations of each other, and has evolved as these expectations have changed over time. The diverse roles played by CSOs have created diverse expectations: civil society as a whole expects more favourable legal, regulatory, and political space to pursue their mission, but organisations involved in service delivery have very different expectations from those primarily engaged in rights-based approaches. The current degree of engagement between government and civil society also varies across levels of the government: operational CSOs are more engaged with local governance institutions than with the national government.

Since the early 1990s both Cambodian and international CSOs, with support from international development aid, have pursued their projects and programmes without much connection to the government. The development efforts by government and CSOs operated almost in parallel. A loose legal framework and regulations helped CSOs. Government institutions were new and grappling with crippled democratic institutions, therefore much government attention was concentrated on rebuilding state institutions. As government was highly dependent on the international development community, adverse relations with CSOs would have been seen as regressive action and could have jeopardised international support to the new political regime. Over the last two decades, however, the state has consolidated its position and institutions and is now more willing to confront challenges to the current political and economic paradigms. The current political regime has been playing the popular legitimacy card within the "electoral democratic" framework very well. Despite several deficits in the democratic framework, a moderate macro-economic stability has given the state both domestic and international legitimacy. This has also made CSOs' challenges of the government difficult. Rejection of CSO criticisms by a legitimately elected government are presented as serving the people of Cambodia.

The recent face-off between CSOs and government over the newly proposed NGO and Associational Laws throws some light on the emerging relationship. The government invited CSOs to comment on several drafts of the new law, which CSOs saw as attempts to tighten government control over CSOs. Currently, with an intervention from the Prime Minister, the draft law is shelved for the time being. But CSOs who were organising ASEAN People's Forum in 2012 faced a lack of cooperation and hostility from the government.

The relationship between government and CSOs involved in service delivery programmes is primarily characterised by cooperation and mutual support. In most cases services are provided to underserved communities, independently of governmental institutions and require interaction only when CSOs need support from line ministries. These CSOs generally feel that government is supportive and cooperative. Other CSOs involved in community development and livelihood generation in the areas of agriculture, livestock, and fisheries have similar impressions. As these interventions promote economic development of poorer communities, government including local officials found them supportive of their development agenda.

Perhaps the most positive relationship is exhibited by group of CSOs which support decentralisation policies and programmes. These CSOs play roles like providing capacity development for newly elected representatives (particularly women elected representatives), extending outreach and mobilisation functions of local governance institutions, facilitating and providing technical support in planning, and so on. The elected representatives particularly from Commune/*Sangkat* and district councils appreciate the information and training support that they receive from these CSOs. Many CSOs also mobilise communities in preparation of development plans which are the responsibility of Commune/*Sangkat* Councils. In the absence of any structured capacity development support from higher tiers of government, CSOs have provided technical support in preparing plans which have in turn enhanced the credibility of the Commune/*Sangkat* Councils. Many Commune/*Sangkat* members who are also members of political parties have testified to the credibility of these CSOs.

CSOs involved in new social accountability practices at the local level often face mixed responses from local governments. CSOs that monitor the functioning of Commune/*Sangkat* Councils or de-concentrated line departments experience opposition from local officials and elected representatives. However, where CSOs have blended monitoring with capacity support to elected representatives a form of constructive engagement is evolving.

Perhaps the most confrontational and mutually castigating relationship exists at present between government and CSOs which work on human rights issues. As many such CSOs also support people's movements against encroachment of land, forest, or water rights, these CSOs are often seen as agents of political opposition. These CSOs must continuously prove their political neutrality. Many feel that the government encroaches on their autonomous space through imposing restrictions.

Relationship with other actors

The formal relationship between CSOs and political societies in Cambodia is in its nascent stage. While each actor is aware of and informally relates to the other, the relationship remains unstructured. CSOs are quite conservative about admitting any relationship with political parties publicly. Given the current dominance of one political party and its aggressive stand on political opposition, CSOs build relations to political parties at some risk. Development-oriented CSOs maintain low key relationships with political parties, but human rights groups are almost underground. Some CSOs interact with elected representatives (for example, Member of Parliament or National Senate Members) as their activities demand. The lack of vibrancy in relations between civil society and the political sector also affects the efficacy of CSO advocacy. In other Asian countries with better relationships, political leaders are helpful in generating awareness and bottom-up demand within their own political institutions for policy reforms.

Currently, CSOs tend to see the relationship with media from an extremely narrow prism. The engagement with both print and electronic media is seen as public relations exercise, mostly to publicise the good work of CSOs. This may in the long-run help civil society to improve its image, but emphasising only one kind of engagement may undermine other potential roles of media in Cambodian society and polity. The independence and autonomy of electronic and vernacular Khmer media is widely questioned as, with very few exceptions, most of the media institutions are aligned with ruling political parties. This is also a reason why some CSOs, particularly those who are politically aware and engaged in governance, democracy, human rights, etc. are very selective about media engagements. As a result the civic and political education function of mainstream media is negligible. Some CSOs have developed their own media for public outreach programmes. Some radio stations (which are also very popular in Cambodia, particularly in rural areas) are more credible in the eyes of CSOs and they have used radio as medium for public education and outreach. Some technologically savvy young CSO activists use social media as platforms for free expression, but their efficacy to influence public or political opinion is yet to evolve.

The relationship between CSOs and academia is also underdeveloped in Cambodia. One reason is that the higher education system remains underdeveloped. Most universities, public or private, do not have well developed research programmes. The quality of curricula and teaching requires many improvements, and engagements with individual academics are also negligible. A few academics join some CSO activities from a philanthropic point of view, but more meaningful and complementary engagements are missing at this moment.

Emerging conclusions: future implications on strategies and capacities of civil society

Political and economic governance issues and the associated challenges of inequity, poverty, and exclusion pose substantive challenges to Cambodian CSOs. The post-conflict reconstruction roles they have been playing for last two decades require radical transformations.

A majority of Cambodian CSOs deliver services with resource support from international donors. As the country becomes economically better off due to its current economic growth, the sources of international funds to CSOs will decline. As has happened in many developing countries, government may want to outsource service delivery functions to private agencies, including CSOs. CSOs with more expertise may have a strategic advantage to assume service delivery functions funded by government. But this choice can make them dependent sub-contractors to the government. The success of CSOs in service delivery programmes may stem from the flexible international support received so far. It is not clear that CSOs will be equally successful if these are funded by government agencies that allow little flexibility or space for innovations.

Cambodia is passing through a critical phase of governance reforms. International donors that support governance reforms may also support CSOs to promote bottom-up, demand-driven approaches. Their appreciation of CSOs will depend on CSO capacities to promote citizen participation in decision-making, foster demand for transparency, and enable government accountability. The ability of CSOs to promote and defend rights and entitlements of citizens would definitely assume greater importance. Assuming these roles may demand new capacities in civil society as well as greater engagement with government. The continuation of historical indifference towards government has to change, with meaningful engagement without losing the autonomy. Some CSOs have started playing these roles and are ready to engage with government on governance reforms, but they are still a minority in Cambodia. CSOs need to develop agility and alliances to manage the combination of support and confrontation needed for such roles.

The current economic growth paradigm led government to provide large-scale economic land concessions to domestic and foreign businesses, displacing residents, undermining livelihoods and denying rights to a sizeable rural population. Recently, there have been increased displacements of urban slums as well. Some of these displaced people, with few alternatives and little hope of justice, have now organised movements and protests. Many others remain unorganised and unheard. Some CSOs offered legal and organisational support but others remain indifferent to such movements. Understandably, CSOs are cautious as the government dislikes criticism and is tempted to see support to these movements as aligning with political opposition. The government is determined to forge ahead with these economic policies and no initiatives to cope with their impacts are in sight, so it is expected that these eruptions of citizen discontent will continue to grow. Can a large section of CSOs afford to remain silent? Should this happen it could create an irreparable chasm between citizens and CSOs. From where will CSOs draw their social legitimacy, if they are seen as irrelevant to the problems of grassroots groups?

Other elements of civil society are disjointed, underdeveloped, and less recognised. Grassroots groups and movements are gradually emerging in Cambodia, but their relationship with CSOs is sporadic and limited. The membership bases of trade unions have grown over the last two decades, but remain restricted to formal economic sector. Relationships between trade unions and intermediary CSOs are negligible, except for a handful CSOs which work on labour-related issues. Relationships between traditional civil society organisations, like Pagoda Associations, and intermediary CSOs are also rudimentary. This lack of connectedness among segments of civil society undercuts the vitality and influence of the sector. For example, the Ministry of Interior published the latest draft of NGO law at the same time that the Ministry of Labour published proposed amendments in trade union laws. While both NGOs and trade unions are part of civil society, they did not forge an alliance to engage with government collectively. Each stood on its own ground, with limited conversation between them. Moving forward, enhanced interconnectedness within civil society would enhance its effectiveness and its political space vis-à-vis government.

CSO engagement with media and academia is also inadequate. Cambodian society would benefit from a more vibrant relationship with both. While the independence of both media and academia from political parties is questionable, CSOs might benefit greatly from building better relationships with these important actors. A trusted relationship between civil society, media, and academia would be useful in opening up political spaces for CSOs on one hand, and in pushing other actors to be more accountable to the other.

Little relationship exists between CSOs and private sector. Although the government adopted a free market economy in 1993, private sector investments have only begun to be important in the last decade. Government concessions to attract domestic and foreign investments are now in place; projections indicate that they will grow in the next decades. CSOs are largely unprepared to assume any role in these developments beyond their current emphasis on criticism and confrontation. The growth of private sector could contribute to poverty reduction provided CSOs could play a meaningful role in partnership. Evidences from many countries indicates that unbridled wealth creation can enhance inequity and injustice, but new forms of social entrepreneurship in partnership between private sector and CSOs can unleash energy and innovations for sustainable and equitable growth as well as more meaningful and effective corporate philanthropies.

Since the early 1990s international aid has contributed to the tremendous growth of CSOs in Cambodia. More recently Cambodian CSOs have started interacting with regional and global processes, so their capacities to appreciate, analyse, and take informed positions will be crucial. A few CSOs have hosted important regional and global events in Cambodia (for example, Open Forum on the International Framework for CSO Development Effectiveness) or participated in global forums (for example, the Fourth High-Level Forum on Aid Effectiveness in Busan), and such capacities need to be spread more widely among other CSOs. An important point for Cambodian CSOs will be defining the national interest from the citizens' point of view; a deeper connection with people and their issues would help to shape their perspectives, analyses, and articulations in global forums.

The big CSOs engaged in service delivery do not face immediate declines in international resources to support their projects. However, CSOs involved in rights-based work do see declines in those resources. Recent trends in international aid indicate that resources for CSOs will continue to flow in Cambodia for the next five years. If Cambodian CSOs are able to draw lessons from countries like the Philippines or Indonesia, where international development aid started declining a decade ago, they will perhaps be better prepared to face resource shortages in the future.

Continued relevance of CSOs in Cambodian society and polity beyond post-conflict reconstruction and service delivery will require critical capacity development in several areas. Given the current tendency toward hegemonic politics and governmental control, CSOs need to develop effective strategies for ensuring checks and balances of power in governance. This would require deeper understanding of the political economy and governance institutions in Cambodia. This enhanced understanding, with well thought out strategies and capacities for engagement, has to be nurtured in the leadership of civil society, particularly among the youth. The demographic advantage cannot automatically provide dividends unless there is thoughtful investment in building new civil society leadership. The capacities for fundamental research, policy analysis, advocacy, and lobbying will be absolutely critical to enhance the relevance of civil society in changing contexts. As new generation Cambodians will have better education, a wider worldview, and greater access to information, civil society needs to embrace this growing intellectual, political, and social capital.

References

Barton, M. 2001. "Empowering a New Civil Society: Pact's Cambodia Community Outreach Project". Accessed September 1, 2012. www.pactcambodia.org/Publications/NGO_Training/empowering_a_new_civil_society.pdf

CCC (The Cooperation Committee for Cambodia). 2012. *CSO Contributions to the Development of Cambodia 2011*. Cambodia: CCC.

Malena, C., and K. Chhim. 2009. *Linking Citizens and the State: An Assessment of Civil Society Contributions to Good Governance in Cambodia*. Cambodia: The World Bank.

Mysliwiec, L. E. 2004. "Envisioning a New Paradigm of Development Cooperation in Cambodia." Phnom Penh: Cambodian Development Resource Institute. Accessed September 15, 2012. www.cdri.org.kh/webdata/download/envisioning.pdf

Nuon, V., and M. Serrano. 2010. "Building Unions in Cambodia: History, Challenges, Strategies." Singapore: Friedrich-Ebert-Stiftung, Office for Regional Cooperation in Asia.

Royal Government of Cambodia. 2004. "The Rectangular Strategy for Growth, Employment, Equity in Cambodia. Accessed September 1, 2012. www.cdccrdb.gov.kh/cdc/aid_management/RGC_Rectangular_Strategy_2004.pdf

Royal Government of Cambodia. 2005. "Strategic Framework for Decentralization and De-Concentration Reforms". Accessed September 1, 2012. www.ncdd.gov.kh/attachments/026_2005%20Strategic%20Framework%20for%20Decentralization%20and%20Deconcentration%20Reforms(Eng).pdf

Royal Government of Cambodia. 2010. "National Program for Sub-National Democratic Development". Accessed September 1, 2012. www.ncdd.gov.kh/en/national-program

World Bank. 2005. "Country Assistance Strategy for the Kingdom of Cambodia." Phnom Penh: The World Bank, Southeast Asia Country Unit, East Asia and Pacific Region.

The authors

Kaustuv Kanti Bandyopadhyay (corresponding author) is the Director of PRIA. He is a researcher and practitioner of capacity building, civil society, democratic governance, social accountability, participatory planning, monitoring, and evaluation. For the last 20 years, he has been associated with universities, academic institutions, and CSOs. He has a PhD in Anthropology.

Thida C. Khus is the Executive Director of SILAKA, Cambodia. She has been a trainer, facilitator, and social worker for over 32 years. Her areas of expertise include capacity building, democratic decentralisation, social accountability, and political leadership development of women. She has served as an advisor on various committees of the Royal Government of Cambodia.

Is civil society in the Southern Cone of Latin America at a crossroad?

Anabel Cruz

If there is one region whose political landscape has dramatically changed in the last 20 years, that region is Latin America. After several decades of dictatorships in many countries, the region is now engaged in strengthening the essential elements of representative democracy. Many events and developments affect civil societies in the region, and citizens and organisations are trying to respond to new or re-emerging challenges with varied forms. This article provides an overview of these events, and civil society responses.

La société civile dans le Cône Sud de l'Amérique latine se trouve-t-elle à un carrefour ?
S'il y a une région dont le paysage politique a connu des changements spectaculaires au cours des 20 dernières années, c'est bien l'Amérique latine. Après plusieurs décennies de dictatures dans de nombreux pays, la région a désormais entrepris de renforcer les éléments essentiels de la démocratie représentative. De nombreux événements et progrès ont une incidence sur les sociétés civiles de la région, et les citoyens et les organisations tentent de répondre aux défis nouveaux ou ré-émergents par des moyens divers. Cet article propose un aperçu de ces événements et des réponses de la société civile.

¿Se encuentra en una encrucijada la sociedad civil del Cono Sur de América Latina?
Si existe en el mundo una región donde los regímenes políticos han cambiado dramáticamente durante los últimos 20 años, esa región es América Latina. Tras varias décadas de dictaduras en numerosos países, la región vive un proceso de fortalecimiento de los aspectos esenciales de la democracia representativa. Las sociedades civiles de la región han sido afectadas por muchos eventos y acontecimientos, por lo cual los ciudadanos y las organizaciones pretenden responder de distintas maneras a los retos nuevos o antiguos. El presente artículo ofrece un bosquejo de dichos eventos y de las respuestas que la sociedad civil les ha dado.

A sociedade civil do Cone Sul da América Latina está em uma encruzilhada?
Se há uma região cujo cenário político tem mudado dramaticamente nos últimos 20 anos, esta região é a América Latina. Após várias décadas de ditadura em vários países, a região está agora engajada em fortalecer os elementos essenciais da democracia representativa. Muitos eventos e desenvolvimentos afetam as sociedades civis da região, e os cidadãos e as organizações estão tentando responder aos novos ou reemergentes desafios de várias formas. Este artigo oferece uma visão geral destes eventos, e as respostas da sociedade civil.

Introduction to the historical political context[1]

The political landscape of Latin America has changed radically over the last two decades. After several decades of dictatorships in many countries, the region is now engaged in strengthening the essential elements of representative democracy. Civil societies, understood as the arena outside of the family, the government, and the market, where people associate to advance common interests, are profoundly affected by events and developments in the region, and citizens and organisations are trying to respond to new or re-emerging challenges.

Currently, all countries in the region have multi-party systems, with periodic, free and fair elections, based on secret balloting and universal suffrage, though weaknesses of some systems have been detected and denounced. But most countries in Latin America endured some form of repressive regime in the twentieth century and suffered violence, violations of human rights, and repression of their civil liberties. This happened both in Central America and in South America as well. Some paradigmatic examples of this include the military regimes in Argentina from 1976 to 1983, in Uruguay from 1973 to 1985, and in Chile from 1973 to 1990 (Barreiro and Cruz 1988). An estimated 30,000 people were killed by the Military Junta that ruled Argentina for 13 years and thousands of prisoners populated the prisons in these three countries. The military dictatorship in Brazil lasted for 21 years, ending in 1985. The dictatorship of the Somoza dynasty ruled Nicaragua for more than 40 years, ending in 1980. Honduras was ruled by a military dictatorship from 1972 to 1980. In El Salvador authoritarian governments employed political repression to maintain power from the 1930s into the 1970s, and in the 1980s the country endured a violent and bloody civil war, which ended with the Chapultepec Peace Accords in 1992.

Thus, when the Berlin Wall fell in 1989, it coincided with the return to democracy in Chile after the 17-year dictatorship of Augusto Pinochet, and with the first steps on the democratic path taken by most Latin American countries. The two decades that have passed since those events have brought back respect for human rights and fundamental freedoms, and laws guaranteeing access to public information have recently been approved in several countries. The free elections and the freedom to run for power have also brought choices of leaders that the citizenry have enthusiastically embraced, and this is an important difference compared with 20 or 30 years ago. As we look at the recent and present Latin American political map, and specifically at the Southern Cone countries, we see that presidents elected during the last two decades include a former union leader, three women (two of them former political prisoners), and a former guerrilla leader. These were unthinkable choices just a few years ago.

Furthermore, during the 1980s and 1990s, the Latin American region experienced a double transition: not only did it start on a new democratic path, but a neoliberal economic model was installed in many countries. This latter was part of the demands of a globalisation process; transitions have been difficult and conflicted. Though democratic processes have moved forward, democracy is still fragile in the region and some authoritarian and non-inclusive tendencies remain.

Civil society groups were crucial in the recovery of democracy in the 1980s and in the construction of a better democracy in the Southern Cone countries of Latin America (Argentina, Chile, Uruguay, and also Brazil). National civil society organisations and citizen groups have been trying to strengthen democracy in region for several decades now using a variety of methods, tactics, and alliances.

Crossroads for civil society

Some common questions for civil society in Chile, Argentina, and Uruguay emerge clearly. Have the advances in democracy-building meant the strengthening of civil society? Has the path to stronger, formal democratic institutions led to the consolidation of civil society? Are there differences between the civil society that made possible the recovery and later transition into democracy, and the citizen organisations active in this new century? The 1980s and 1990s produced specific social organisations in these countries; NGOs were active and significant during those two decades and became stronger under the protection of international cooperation. Those organisational forms are not entirely relevant any longer, and new forms of civil society are emerging. More informal movements, independent actions, and street protests based mainly upon voluntary forces represent the most interesting, challenging, and unexpected new practices that are emerging in the activities of civil society in the region.

New and expanded spaces

In Latin America we are witnessing the creation of new innovative and expanded spaces for civil society, with important amplification and expansion of rights and claims of rights by social collectives and social movements. The uprising of the student movement in Chile in 2011 focused on the right to education, pointing an accusative finger against a political system grounded in the neoliberal logic since the dictatorship years (1973–90). The women's movement and other groups in Uruguay are demanded sexual and reproductive rights and the legalisation of abortion. The lesbian, gay, bisexual, and transgender (LGBT) movement in Argentina exemplified new civil society formations that advance rights not included in the first waves of citizenship-building and absent at the time of the democratic transitions. Their campaigns laid the foundation for an expanded system of rights that recognise and valorise diverse sexual orientations.

These campaigns focused on central demands but also sometimes raised other important questions in the course of their activities. The right to the use of public spaces, for example, was raised in Chile by the student movement. These examples represent social movements that demand not only reforms but radical changes, expanding the concept of rights and fields of possibility.

Resources for civil society

Funding from democratic governments in the North during the 1970s and 1980s was crucial in the recovery of democracy in Argentina, Uruguay, and Chile in the 1980s and in its subsequent consolidation. It has been broadly recognised (Berro, Barreiro, and Cruz 1996) that international cooperation played a crucial role in the recovery of the democratic course: human rights groups, women's groups, independent research centres, and civic groups received support and funding that allowed them to have a voice, to exert pressure, and finally to contribute to the reinvigoration of formal democracy. Key civil society groups were supported by international donors, and that support made a positive contribution not only in financial terms but also in breaking the local and international isolation in which groups operated.

But the trends in international cooperation have changed in the last two decades in Latin America. Funding from several sources has moved from supporting participatory mechanisms of citizen engagement to the building and strengthening of formal democratic institutions (growth of electoral systems, advancement of judiciary systems, etc.). In many cases, funding for civil society organisations has been reshaped to flow through governmental offices and special "windows" managed by ministerial departments or similar agencies. Unfortunately, Latin America has ceased to be a priority for international cooperation agencies, for countries in the North, for progressive foundations, and for development NGOs. Some countries are experiencing an end to international cooperation resources: Argentina, Uruguay, and Chile, for instance, now qualify as upper-middle income countries that are not eligible to receive funds. The implications for civil society organisations, especially in Uruguay and Chile, have been manifold. Many groups have become heavily dependent on national and local government funding, a reality filled with challenges. In Uruguay a recent study showed that at least 50 per cent of the budget of NGOs comes from government; in Chile it is an even higher percentage (ICD 2011). This dependence can undermine the autonomy of the organisations. When sustainability is in danger, some organisations may avoid confrontation in order to survive, but lose sight of their missions in the process.

In such a complex situation, with still fragile democracies coexisting with the consequences of economic neoliberal theories and important sectors of the population suffering from increasing poverty, civil society organisations face diverse challenges. The panorama, of course, is not homogenous: in some countries or sub-regions, civil society organisations are have been immobilised for periods, with very low capacity for critical evaluation. In other cases a process of phagocytising, or co-optation, has taken place, so civil society organisations implement government responsibilities, becoming mere executors of already-designed policies rather than innovators or guardians of the public interest.

Few questions are more controversial and complex than democracy in Latin America. To the classic discussions related to formal and substantive democracy is added the demand for deepening democratisation processes through more citizen participation. This perspective includes questions of legal and social justice but also focuses on solidarity practices, citizen participation, and quality of democratic process. Funders from countries in the North, both governments and private organisations, and even indigenous philanthropy are being called to reflect on how they can contribute to active citizenship and democratic institutions (Ibero-American General Secretariat 2011).

From protest to proposals

Recent years have seen citizen uprisings in Latin America and the consolidation of new forms of civic participation. In Chile, students developed street mobilisations and huge protests, including marches and demonstrations, occupations and hunger strikes, and many forms of creative expression and virtual activism for their campaign for education reform. The student movement protests in Chile have created the largest and most participatory mobilisation since the restoration of democracy in 1990. The sexual and reproductive rights movement in Uruguay and the LGBT movement Argentina have also used street demonstrations and rallies, though on a smaller scale. Protests were not ends in themselves but a way to convey support for proposed legislation – such as the laws to legalise abortion in Uruguay or equal marriage rights and gender identity in Argentina – and a means for proposing alternative solutions, such as a plebiscite on education in Chile. Unlike many protests, these movements focused on large scale reforms and policy changes backed by broad coalitions rather than opposition to concerns that affected narrow constituencies.

Enriching the public agenda

In recent years, social and rights movements have been placing new topics on the public agenda and in the public scenario. Transparency and accountability issues, anticorruption, rights of minorities, and several other matters have been pushed by citizens groups in the last two decades. Women's movements have changed how mass media deals with sexual and reproductive rights, broadening the issues and topics they cover and changing the language used by the media and inside the movements themselves. Sexual rights, sexual diversity, or women's rights to control their bodies were not discussed publicly just two decades ago; it was unthinkable that they could become the glue that linked different actors in a social movement for rights. The path that led to the approval of the Gender Equity Law in 2012 and the Equal Marriage Act in Argentina started only seven years ago with a discussion about the location for prostitutes in the city of Buenos Aires. The women's movement in Uruguay translated a defeat into an expansion of their alliances after a Congress-approved bill was vetoed by the President. They changed the name of their coalition from *movement for sexual and reproductive rights* to *coalition for the legalisation of abortion*. As for the Chilean student movement, it succeeded in forcing former taboo issues – such as tax reform – onto the political agenda.

New organisations and channels of participation

These initiatives in the Southern Cone in Latin America reveal that the typology and morphology of civil society are changing at both national and regional levels. New organisations and coalitions have been created in recent years, revealing new forms of governance and hybrids of traditional civil society formations. Organisations such as the Coordination for Legal Abortion in Uruguay, the Argentina LGBT Federation, the students' federations in Chile, and the chapters of the regional organisation TECHO were all founded during the past decade. Some older organisations were revitalised by their participation in these movements and by new challenges.

Often in Latin American society it is said that young people are not interested in altruistic causes or in collective projects, because they are steeped in the individualism of the cultural model. Apathy and disinterest in political parties and a lack of knowledge of political issues have been debated in the press, explored in academic studies and discussed by political parties in many countries. Even when progressive parties and coalitions are in power in nine countries, youth participation has not increased. In several countries, the estrangement of young people from partisan politics has increased over the last 20 years. Some suggest that uninterested young people and unattractive politics are important reasons for this estrangement (Gonzalez-Vaillant 2012).

In the context of low political participation of young people in social and political organisations in most countries in Latin America, the massive youth participation in the student movement in Chile is important. Likewise, the regional movement TECHO, a youth-led non-profit organisation that works in 19 countries in Latin America and the Caribbean to promote cooperation among poor families and youth volunteers to overcome poverty in slums. TECHO have combined work for social development with advancing the right to dignified housing by convening more than half a million young volunteers, mostly university students, in recent years. TECHO stress their mission to transform volunteers as well as provide specific assistance, so they want to go beyond traditional forms of volunteering. The high impact and strong convening power of this regional movement has attracted the attention of other organisations, state actors and private companies.

Alliances and partnerships

The new coalitions and organisations that are leading these social movements included examples of intense interaction with different actors and organisational allies as well as individual supporters. As very important processes in Chile, Uruguay, and Argentina, the student movement, the women's movement, and the LGBT movement generated pivotal alliances with trade unions. Alliances with trade unions are not common in civil societies in the south of Latin America. The intergenerational alliance of the Chilean student movement, which mobilised parents, grandparents, and younger siblings of participants, broadened demands from student-specific to society-wide concerns.

Articulations and interactions with political parties vary from country to country, depending on the historic moment. While in Chile there might have been a confrontation between the student movement and the political society, in Uruguay there is a clear fracture inside the political society and within political parties. Relationships with governments must be highlighted as well as those with political parties, especially in the countries with progressive and leftist political parties in power. It is important to analyse the possible deactivation of the social movement, the co-optation of the demands, or just the possibility of the government adopting the demands. The implications for civil society in countries with leftist governments (or self-proclaimed progressive governments) are multiple. Argentina, Uruguay, Brazil, Peru, El Salvador, Nicaragua, Venezuela, and Chile (during the 20 years from 1990 to 2010) can be included in this group. In Chile the *Concertacion* of Parties for Democracy ruled for 20 years, and even if the first signs of discontent were expressed by students during the administration of Michelle Bachelet, the real mobilisation occurred during the current administration, and its duration has surprised many analysts. In Argentina and Uruguay, civil society organisations have difficulty exerting influence when the government takes up its flagship causes (Antía 2010) and when governments provide much of civil society's funding.

Outreach and impact

More and more often, social movements are becoming aware of the importance of public opinion and mass communication. The student movement in Chile, the women's movement in Uruguay, and the movement for equal marriage in Argentina all made excellent use of new forms of communication, with innovative use of colour, sound, movement, and imagination. These innovations helped to capture the attention and sympathies of the public. This support translated into a high turnout at demonstrations and also into a strong social consensus and backing on reform proposals. In September 2011, according to public opinion polls, 89 per cent of the Chilean population expressed support for students' demands. Public opinion has also become supportive of the demands of legalisation of abortion in Uruguay, but as part of a much slower process (51 per cent support from the population, which has grown from 45 per cent over the last five years) (Bottinelli 2011, 77–127).

These new forms of social organisation transcend the borders of their respective communities, cities, and countries to have an important regional impact. The Chilean student movement received solidarity expressions from all over the world. Actions are more global and less local and national initiatives can become international issues. A movement born in one country and context can become a regional initiative: TECHO is now present in most countries in Latin America, and its founder has moved to Africa with the aim of launching a similar process there.

Last but not least, the measure of failure or success is an important question while analysing the crossroads for civil society in Latin America. In each specific context it is important to consider concrete and tangible results in terms of products (such as housing for poor populations), achievements in terms of legislation approved (such as the Equal Marriage Act in Argentina), and more fundamental changes in paradigms (such as the irreversible changes in attitude toward education and equality in Chile). These new forms of social organisation are transcending the borders of their respective communities, cities, and countries, to have important regional and global impacts. They are also demonstrating that measuring success or failure is not a simple or linear task but a very complex endeavour.

Note

1. This overview is extensively based on the results of the Crossroads studies conducted in Latin America: the Chilean Student Movement in Chile; the Movement for Legalizing Abortion in Uruguay; the LGBT movement in Argentina; and the Regional Movement TECHO.

References

Antía, F. 2010. "Políticas sociales y desarrollo. Los desafíos para las izquierdas de Chile y Uruguay" ["Social and development policies. The challenges for the left in Chile and Uruguay"]. In *Las izquierdas latinoamericanas*, edited by P. Alegre et al., 77–127. Buenos Aires: Consejo Latinoamericano de Ciencias Sociales (CLACSO).

Barreiro, F., and A. Cruz. 1988. *La dificultad de ser. Organizaciones No Gubernamentales en el Uruguay de hoy* [The difficulty of being. NGOs in Uruguay today]. El desafío de la democracia. Montevideo: Instituto de Comunicación y Desarrollo (ICD), Fundación de Cultura Universitaria.

Berro, M., F. Barreiro, and A. Cruz. 1996. *América Latina y la Cooperación Internacional* [Latin America and International Cooperation]. Montevideo: ICD.

Bottinelli, O. 2011. *Aborto: la opinión pública en los últimos 20 años* [Abortion: public opinion in the last 20 years]. Montevideo: MYSU.

Gonzalez-Vaillant, G. 2012. "The Politics of Temporality: An Analysis of Leftist Youth Politics and Generational Contention." *Social Movement Studies: Journal of Social, Cultural and Political Protest* 11: 3–4. DOI:10.1080/14742837.2012.723367

ICD (Instituto de Comunicación y Desarrollo). 2011. *Vamos Andando. Las organizaciones de la sociedad civil rendimos cuentas* [We walk: Civil society organizations are accountable]. Montevideo: ICD.

Ibero-American General Secretariat. 2011. *Report on South-South Cooperation in Ibero-America 2011.* Madrid: SEGIB.

The author

Anabel Cruz is the Founder Director of the Communication and Development Institute (ICD) of Uruguay, a civil society organisation and private centre, with more than 26 years of experience in strengthening civil society and citizen participation, by means of research, training, capacity building, and dissemination of information activities in Latin America and worldwide. She has published several books on civil society and development and her articles have appeared in journals in Latin America, the United States, and Europe. She was the Chair of the Board of CIVICUS: World Alliance for Citizen Participation 2006–10 and is currently the Co-coordinator of the Regional Initiative "Rendir Cuentas," the Chair of the Partnership for Transparency Fund, PTF, and a Board member of Humanitarian Accountability Partnership, HAP International.

Protest and proposal, participation and representation: the Chilean student movement, 2011–12

Inés M. Pousadela

This study examines the student movement in Chile, which brought about the largest mobilisations since the restoration of democracy in the early 1990s. The students protested the so-called "education apartheid" created by the Pinochet regime and preserved by all democratic governments ever since, in which the formerly public education system had been largely replaced by a more massive but also expensive, highly stratified one, financed by private debt. The students' demands for free and high-quality education for all gained the support of broad sectors of civil society, which was expressed both in opinion polls and in massive participation in demonstrations.

Protestation et proposition, participation et représentation : le mouvement des étudiants chiliens, 2011-12
Cette étude examine le mouvement de contestation estudiantin au Chili, qui a donné lieu aux plus importantes mobilisations depuis le rétablissement de la démocratie au Chili au début des années 1990. Les étudiants ont protesté contre ce qu'ils ont appelé l'« apartheid de l'éducation » créé par le régime Pinochet et préservé par tous les gouvernements démocratiques lui ayant succédé, dans le cadre duquel le système d'éducation jusque-là public avait été en majeure partie remplacé par un système plus massif mais aussi coûteux et très stratifié, et financé par la dette privée. Les exigences des étudiants concernant une éducation de qualité pour tous ont obtenu le soutien des grands secteurs de la société civile, ce qui a été exprimé tant dans le cadre de sondages d'opinion que dans la participation massive aux manifestations.

Protesta y propuesta, participación y representación: el movimiento estudiantil chileno, 2011-2012
El presente artículo examina el movimiento estudiantil de Chile, impulsor de las movilizaciones más grandes que han tenido lugar desde la restauración de la democracia a principios de los años noventa. Los estudiantes protestaron en contra del llamado "apartheid educativo" creado por el régimen de Pinochet y sostenido por los siguientes gobiernos democráticos. Dicho apartheid significó la eliminación del anterior sistema educativo público y, en gran medida, su reemplazo por un sistema de mayor alcance pero también más caro y más estratificado, financiado por la deuda privada. Las demandas de los estudiantes para instaurar una educación gratuita de alta calidad fueron respaldadas por amplios sectores de la sociedad civil, lo cual se expresó tanto en las encuestas de opinión como en su masiva participación en las manifestaciones.

Protesto e proposta, participação e representação: o movimento estudantil chileno, 2011–12
Este estudo examina o movimento estudantil no Chile, que promoveu as maiores mobilizações desde a restauração da democracia no início da década de 1990. Os estudantes protestaram contra o chamado "apartheid educacional" criado pelo regime de Pinochet e preservado por todos governos democráticos desde então, no qual o sistema educacional público antigo havia sido em grande parte substituído por um sistema mais massivo mas também caro, altamente estratificado, financiado pela dívida privada. As reivindicações dos estudantes por uma educação gratuita e de alta qualidade para todos receberam apoio de amplos setores da sociedade civil, que foi expressado tanto nas pesquisas de opinião quanto na participação massiva nas demonstrações.

Introduction

This study explores the changes that occurred in the organisational forms of social protest and their articulation with the political system, through the analysis of the recent mobilisations lead by the Chilean students. This process of mobilisation – which included demonstrations, marches, occupations, hunger strikes, diverse creative expressions, virtual activism, scattered violence, and harsh police repression – started in April 2011 and soon became the largest one since the restoration of democracy in the early 1990s.

The student protest pointed an accusative finger against the whole political system, inasmuch as the neoliberal logic governing the Chilean education system since the dictatorship years (1973–90) remained basically unchanged not just by the present right-wing government but also by its center-left predecessors. Since many of the basic principles that govern the Chilean socio-economic model have constitutional status, the students' demands logically led to the demand of a reform of the Constitution. Far from being an eccentricity of the student movement, this demand was eventually shared by 75 per cent of Chileans (CERC 2011b).

Under the Pinochet dictatorship, basic education was transferred from national to local governments, free public university was abolished, the University of Chile was dismembered in regional units, and the requisites for setting up and running private educational institutions were drastically reduced. As a result of the ensuing boom, the system modified its composition: nowadays, 60 per cent of elementary and secondary school students are enrolled in private schools, most of them subsidised, whereas only a handful attend elitist private schools. The decentralisation of education increased inequalities, since quality came to depend upon the resources of each municipality. Thus, despite Chilean education ranking higher in quality than that of other countries of the region, it also ranks highest in segregation and is therefore described as a system of "education apartheid".

With the creation of approximately 30 new private universities following liberalisation, access to higher education rapidly ballooned, increasing from 250,000 to about a million students between 1990 and 2010. Nonetheless, university education access rates remained extremely low among lower income groups. At present, 60 per cent of students – among them the poorest, who come from the worst high schools and therefore fare worse in admission exams – attend private institutions with poor academics. Public universities receive scarce government funds and are the most expensive in the region. Consequently, 70 per cent of university students pay for their studies by taking long-term bank loans that will consume a high proportion of their future earnings.

By emphasising the economic and political elements (and their intersections and complicities) that have shaped and deepened social inequality and exclusion and affected society as a whole, the student movement managed to attract growing public opinion support. Indeed, far from being perceived as narrowly sectorial, the students' demands for free and high-quality education for all gained the backing of broad sectors of civil society, both in their virtual expression and through their actual presence in demonstrations – which, as a result, became the most massive of the current democratic period. Yet, frequent violence was reported and mobilisations were systematically repressed.

This article is based upon research involving both secondary sources and data from interviews. All online editions of the Chilean newspaper *La Tercera*, and to a lesser extent other newspapers and magazines as well, were revised from April 2011 to April 2012 in order to establish a chronology of events and to collect data for a description of the movement's actions and interventions, as well as discourse fragments allowing for the exploration of conceptual debates and other controversies.[1] Likewise, web pages, Facebook profiles, tweets, and recorded videos posted in YouTube by the students themselves were examined. Additionally, a focus group was conducted with university students from Concepción, and data was compiled from short interviews administered during demonstrations in Santiago de Chile, as well as from a few in-depth interviews with mobilised students and qualified informants. Overall, the testimonies of 40 participants were included in our analysis.

Have students become a new political subject? How novel are their public interventions, and in what ways? To what extent has their protest been accompanied by specific proposals the implementation of which could be used as a measure of success? How have they interacted with those in a position to make that happen? How effective have they been – or could they ever be – in achieving change in the desired direction? In the following pages, these questions are approached through the description and analysis of the students' organisations and leadership; their forms of political action and expression; their media impact and their relationship with public opinion; their discursive interventions within what became a true cultural battle; their democratisation demands, their criticisms to representation and their forms of legitimisation; and their links with politicians, political parties, and the government.

The chain of events

The 2011 mobilisation grew from a series of earlier protests by secondary students that had put an end to a decade of passivity following the gradual, negotiated transition to democracy. The present movement inherited from earlier mobilisations the demand for structural change in the education system as well as scepticism about leaving the solution "in the hands of politicians".

The process began in April with a student strike at the Central University of Chile, one of the oldest private schools, against a potential sale to a for-profit holding. Within two weeks the campaign had managed to articulate a more general demand to enforce the existing ban on profit in higher education and strengthen government regulation of private university operations. In late April the CONFECH (the national confederation of university students) convened the first national mobilisation. A second mobilisation in May, endorsed by the National Workers Council and the Teachers Federation, summoned 15,000 people in Santiago, as well as smaller numbers in other cities. These events were followed by clashes between the national police and violent masked youths known as *encapuchados.*

Deeming official responses to their demands insufficient, students called for a new mobilisation on 26 May and then for a national strike on 1 June, which brought out 20,000 students in Santiago and smaller crowds in many other cities. Subsequently, a number of universities were occupied by students and/or suspended courses until further notice. In early June secondary school students began occupying their schools; by the end of the month six hundred secondary schools had been taken.

Also in June, increasing numbers of students and other citizens marched in major cities, with a handful engaging in violent clashes with authorities. Government proposals were rejected as *papering over the cracks* rather than bringing about the needed fundamental change. At the end of the month a so-called "social strike" mobilised more than 80,000 people in Santiago and twice as many nationwide. A presidential proposal for changes in the educational system was rejected in July and a third national strike mobilised many social organisations and unions as well as students. Vandalism at the end of demonstrations was harshly suppressed by the police; television reports showed violence was triggered by provocations by plainclothes police posing as students.

Further negotiations among a changing cast of ministers and national students' and teachers' organisations failed to produce agreement and were followed by calls for another national strike in mid-August. A series of demonstrations prior to that date mobilised secondary school students, parents, and teachers. The national strike was endorsed by the Unitary Workers' Central and the Confederation of Copper Workers, among other labour federations and social organisations. The march in Santiago, attended by between 100,000 and 150,000 people, was mostly peaceful, although it was followed by disturbances by masked *encapuchados* that were suppressed by the police.

During the following days the government offered to establish a working forum with student leaders but students rejected the very idea of deals made behind the people's backs. Meanwhile, the Teachers' Federation mobilised citizens in a so-called "civic plebiscite" and later presented its results to the President. On 17 August the Education Minister announced an improved proposal that the students rejected as incomplete, vague, and a tool for perpetuating student indebtedness. Despite rain and snow, the next day a new mobilisation summoned more than 100,000 people in Santiago, and more than 230,000 in the country as a whole. Three days later a so-called "family demonstration" in Santiago brought out a crowd that its organisers estimated at almost a million people.

Soon after, a two-day National Citizen Strike summoned by social, union, and student organisations – and joined even by a few political parties – was attended by approximately 400,000 people. Right after noon vandalism and violence spread despite the demonstrators' efforts to stop them. The day ended with 1,394 arrests and the death of a 16-year-old hit by a bullet fired from a passing police car. On the following day, the President personally summoned students, parents, teachers, and deans to sit down and talk. The organisations of students, teachers, and deans agreed to meet the President and his Minister of Education at La Moneda, where they received a proposal for a dialogue table – which students rejected on the grounds that it did not offer minimal guarantees for a negotiation on an equal footing to take place.

In early September thousands of people dressed in black participated in "silent marches" through Santiago and other cities. Soon after, student and teacher organisations came up with four conditions conducive to dialogue: deferring debate of the education bills already sent to Congress, postponing the end of the first semester, broadcasting the debates of the dialogue tables, and interrupting funding to universities that made profits. The Minister of Education rejected the first two conditions and partially agreed to the other two. Students and teachers, however, rejected his counter-proposal and organised a march along the Alameda – including performances by well-known artists – that brought together 180,000 people in Santiago (and 300,000 nationwide). Meanwhile, at the UN General Assembly the President described the student movement as mobilised by "*a noble, big, beautiful, and legitimate cause*" that his government shared (UN 2011). His words were widely criticised as validating abroad the demands that were denied at home.

In October the President and the Minister of the Interior announced a bill that would impose jail sentences on those responsible for school occupations. Two days later almost 40 demonstrators from technical-vocational schools were arrested. The following day, university and secondary student representatives decided to abandon the dialogue with the government since, they argued, the latter was clearly not willing to move forward with substantive issues.

On 6 October, a new march along the Alameda – not authorised due to disagreements about the prospective route – was dispersed with water cannons and tear gas. Police installed fences and positioned themselves before dawn to prevent demonstrators from gathering. Later in October and November, two-day national strikes were organised. Within Congress an intense debate took place about the Education portion of the 2012 Budget Law, which was passed in late November.

Between October and November universities held assemblies, plebiscites, referendums, and consultations to decide whether to begin the second semester. Some decided to return to classes with "protected schedules" in order to carry out assemblies and demonstrations; others decided to remain on strike. In some cases deans tried to force students to return to school despite their decision to stay on strike. Towards the end of November, nevertheless, most universities were having classes again. This change in strategy was justified on the grounds that there was no incompatibility between mobilisation and class attendance, and that achieving free education for all was a long-term task.

Among secondary students, occupations lasted longer: the last occupied schools were returned to their authorities in January 2012. The expulsion of students considered to be agitators or suspected of disorderly behaviour during occupations proved hard to resolve and many re-incorporations required court orders. This process continued well into 2012.

As the year ended, the campaigners emphasised the need for a new phase focused on long-term strategies. The last few activities of the year included a national protest against the admission exam for the higher education system, which was criticised for being a test of students' socio-economic status rather than their talents. On 29 December a new Minister of Education was appointed. Mobilisations, mostly by high school students, resumed in April 2012 and continued throughout the year.

Actors, leaders, and organisations

After decades of conformism, the youth came massively to the forefront of the political stage. Among the student leaders stood out Camila Vallejo, a beautiful, charismatic, outspoken 23-year-old undergraduate student and Communist Party activist who had just become the president of the Students' Federation of the University of Chile (FECH). By the end of 2011 35 per cent of Chileans thought of her as the "*person of the year*" (Cooperativa, Imaginacción, and UTFSM, 2011b); others, however, considered her to be the communists' puppet. Other prominent student leaders were Giorgio Jackson and Camilo Ballesteros, presidents of the student federations of the Chilean Catholic University (FEUC) and the University of Santiago de Chile (FEUSACH), respectively. Their successors as well as other leaders within those federations also played important roles at various points of the process.

Vallejo, Jackson, and Ballesteros were Executive Board members of the CONFECH, the national organisation of the student federations of around thirty public and older private universities. As a confederation, the CONFECH made its decisions in assemblies to which member federations sent their representatives to convey their votes (with a mandate derived from discussions and votes within each federation). Secondary education students, in turn, were grouped in two confederations, CONES and ACES. They also made their decisions in assemblies; due to the larger number of schools, their processes tended to be slower than those of university students.

Another important actor was the Teachers Federation, led by another Communist Party member who was not very popular with the wider public and whose positions were usually in agreement with those of the students. Interviewees typically emphasise the importance of "*the unity of youth and teachers, the fact that they all have stuck together regardless of each person's political ideas*" (Female high school student, interview). As put by a teacher, "*the majority of the teachers support the movement. One would need to be an idiot not to*" (Female high school teacher, interview). Another important organisational actor was the council of university deans (CRUCH), which usually had more distant and conflictive relations with the student movement.

Finally, students' parents and guardians also played a role, especially regarding the denunciation of abuses of authority and police violence during the evictions of occupied schools. Some observers argued that the unprecedented reach of the movement was clearly visible in the transformation of the role played by such organisations, hitherto organised from above and devoted to "*charity fairs, little parties*", and now unprecedentedly engaged in political debate and showing support for their mobilised children (Male technical-vocational schoolteacher, interview).

The student movement was far from monolithic. Right from the start differences became apparent between university and secondary school students. While the former exercised a clear leadership role, the latter provided drive and radicalism and risked and lost more in the process. This disparity was the cause of bitter disputes, but it was also the source of productive tensions:

> "Without secondary students this movement is nothing ... Because they are the most radical ones ... they opened the schools to the public and invited people to go in and participate in discussions ... They did it so well that the idea gained wide acceptance among the public. ... [It was] a very good mix, the radicalism of secondary students with the support of university students generated a strong movement." (Male university teacher and PhD student, interview)

There were tensions among both secondary and university students as well. The two main organisations of secondary students represented two defined types of schools: on one hand, the more traditional, elitist ones (CONES); on the other, the less privileged ones (ACES). The government preferred to deal with the former. In turn, the CONFECH only included a limited number of university federations and struggled over whether to admit new ones. Besides, its decision rules gave a vote to each member federation regardless of size and so conferred veto power to a few small federations dominated by extreme-left groups.

Relations were also tense with the organisation representing university deans. The deans often agreed with the official view of student intransigence. Three particularly divisive issues were tuition costs, the internal democratisation of universities, and the decision whether to resume classes or remain on strike. Student leaders usually reminded deans that students were the driving force of the movement and that they would move forward with or without them.

The government developed strategies to widen these cracks with separate initiatives and claims of ongoing conversations and progress with each group within the movement. For example, they decided to engage first with university students and teachers and soon after declared discussions "*well underway*" to keep secondary students in check (*La Tercera*, July 12, 2011). By late July, the Teachers Federation, CONFECH, and CONES joined forces to meet the government together and developed the Social Agreement for Education. The ACES was initially not included, which exacerbated their fears of being sold out. The government also attempted, somewhat more successfully, to co-opt the deans and split them from the students.

The repertoire of actions

Strikes and massive demonstrations were central to the process, but the students employed a remarkably wider repertoire of strategies for the public presentation of demands. The pages that follow illustrate them, their colourful atmosphere, and the creativity and imagination invested in expressing discontent.

Massive demonstrations and public opinion

The student movement had more drawing power than any other mobilisation since the restoration of democracy. Demonstrations grew as they turned into citizen and family marches, uniting entire families. They were also the entryway to the movement for many people without prior political experience. Observers typically described them as "*parties*" and "*carnivals*".

The power held by numbers was reflected by regular disputes about attendance estimates. As a student leader explained, "*the political class is so self-centred that they won't even notice we are there if we don't succeed in calling 200,000 people out to the streets*".

The epicentre of the movement was in the capital city, Santiago de Chile, but there were important mobilisations in many other cities as well. Over time, public opinion became increasingly aware of and favourable to the students and their demands. According to CERC (2011a; 2011b), the government's rate of support fell from 35 per cent in May to 22 per cent in August, while 89 per cent of public opinion supported the students' demands. In December 82 per cent declared the students' demands to be right. However, approval of the way mobilisations had been carried out dropped from 52 per cent in August to 38 per cent in October, even though most people did not deem the students responsible for the ensuing violence. Moreover, although most of the people thought that the government should authorise the marches, only 18 per cent approved of unauthorised marches and 37 per cent approved of occupations (CEP 2011).

Occupation of educational institutions and other sites

The occupation of educational institutions was generally decided by votes within each student centre or federation. Occupations could involve from dozens to thousands of students; many – but not all – were peaceful. Numerous occupations included "cultural actions" staged to communicate demands. Evictions were sometimes peaceful and oftentimes violent, involving physical damage and arrests. They might occur after hours, days, weeks, or even months of occupation; some buildings were later reoccupied. Public authorities reacted in a variety of ways: some university deans tried to prevent occupations by locking the school doors. Similarly, some mayors were more prone than others to request police intervention, and a few of them actually joined the protests to claim more resources for their school districts.

Despite the harsh public controversies that it eventually raised, students saw occupations and strikes as some of the best means available for them to exert pressure effectively. While many of them recognised the costs of losing a year of school, they still believed that the authorities would pay little attention to demands less emphatically expressed.

The concept of occupation was also extended to symbolic takeovers of various places. These were frequently improvised, and often resulted in police evictions. Favourite targets of these occupations were government offices, political parties' headquarters, and television channels.

Education Ministry offices, for instance, were occupied in several regions, as were the headquarters of several political parties. Television stations were taken over to denounce the so-called "communication barrier" created by the media. The National Senate was occupied in order to demand a plebiscite, and the offices of UNICEF were taken by secondary education students to alert other countries about the Chilean situation.

Pot banging

Pot-banging protests have a long history in Chile. They were used for the first time in the 1970s against the Allende government and reappeared in the 1980s against Pinochet's dictatorship. Since democracy was restored in 1990, pot banging was not used until August 2011, when Camila Vallejo called on Twitter for a pot-banging protest to condemn the repression of demonstrations. Thousands of people banged their pots and pans from balconies, front doors, and windows that very night, thus evading the government prohibition of street protest after the disorders. From then on, the student movement repeatedly called for pot-banging protests to accompany national strikes or to denounce repression. This form of protest was used both independently and together with other forms of political expression, such as candle nights.

Hunger strikes

Hunger strikes were resorted to exclusively by secondary education students. The most prominent one took place in the Buin district, where a strike that began with eight students spread to 40 and generated visits of opposition political leaders, officials from UNICEF, and the Human Rights Institute – but not from the Minister of Education. After much highly-publicised debate, the strike was eventually called off on day 37. Long hunger strikes were also launched elsewhere, sometimes with parents participating in them. In September the government urged all remaining hunger strikers to call off the strikes; the last ones finally stopped after 71 days.

Exposures

The strategy of exposure was first used to unmask, with noisy demonstrations at their homes or workplaces, the Pinochet-era repressors and torturers who had remained free and anonymous. In the absence of judicial sanctions, its aim was to arouse social punishment as a result of knowledge of the target's past. These actions were swift and collective so as to avoid individual punishment. The student movement broadened the reach of exposures to individuals seen as violators of the right to education, including mayors who ordered repression of the movement.

Theatrical staging

Massive demonstrations routinely culminated in cultural events – usually concerts – featuring popular artists. Marches generally included percussion samba, elaborate costumes, painted faces, live music and dancing, puppets, artistic installations, and ingenious slogans. But in all cases the mass demonstration remained the main event.

Other protests, often organised through social networks, were instead *defined* by theatrical staging. Among the most original and eye-catching ones were flashmobs, choreographies performed by large groups in costumes and make-up. Among many other creative events were the "symbolic suicides" for education, either performed by the students themselves or simulated with puppets representing hanged students; the artificial beach set up in a central square by 500 secondary students in swimsuits and sunglasses to mock the change of date of their winter vacation; the football match played between Education and Profit; and, in late December, the installation of a giant mailbox in the Plaza de Armas for depositing "Christmas wishes" for education.

Other staging events included walks from distant cities to Santiago or the National Congress in Valparaíso. The longest such walk was a 500-kilometre journey from Concepción to the nation's capital, where students delivered a letter to the President. In June, runners carried out the "1800 hours for Education" relay race around the presidential palace for 75 days non-stop.

New technologies and social networks

These protests were set apart by their intensive use of communication technologies. Hundreds of photographs and videos of the protests were taken by the protagonists themselves. They were posted online and endlessly reposted, discussed, and reinterpreted in blogs and social networks. So it is not surprising that opinion surveys found high degrees of citizen knowledge of the demands and actions of the student movement.

The social networks were essential tools for organisation and coordination, means of communication, and sources of information. The Internet was both a protest scenario and a combat weapon. The website Yodebo.cl, where students published the amounts of their university debts, enabled their protest. The hacker group Anonymous disabled several Ministry of Education websites. Social networks became verbal battlefields that included the exchange of insults, uncivil behaviour, threats, and harsh controversy.

The Internet, and social networks in particular, was also fundamental to international visibility. Virtual contacts resulted in valuable international and transnational support. They also yielded activities such as the "continental demonstration" on 24 November, organised jointly through the social networks by Chilean and Colombian students; and the European tour by CONFECH leaders, which included press conferences, dozens of meetings, and their participation in international forums. Camila Vallejo retrospectively noted that this trip helped them understand that "*our demands are not utopia*", as they were a reality in other parts of the world (*La Tercera*, October 19, 2011).

Points and counterpoints

At the heart of the student movement was a concern about the nature of Chilean society and its failure to live up to the promise of democracy after Pinochet. These concerns gave rise to abundant discourse and a series of debates with the government and within society itself.

The definition of education as a good

At the root of the struggle lie differing concepts of education and of the relationship between society and the state. The student movement defined education as a right of all citizens rather than a privilege for a few. It also saw it as a public good – or, as one student leader put it, as a "*platform for fairness*" and the "*integrative core of our society*" (Giorgio Jackson in *La Tercera*, July 6, 2011, and August 18, 2011). If education is a right and a public good, they reasoned, then the government has a duty to protect and advance it.

On the opposite side, as stated by President Piñera, education was understood both as an "*investment*" and a "*commodity*" and, under the Chilean conditions, as a luxury rather than a mass consumer good. Education was therefore placed within the logic of the market: its delivery, it was argued, should be left to the interactions between supply and demand, and profit should be embraced as its driving force.

Thus, while the government promised, at the most, to control and regulate profit in education, the student movement sought to ban it altogether (or, more precisely, to achieve enforcement of the legal prohibition already in force) because, they argued, it inevitably distracted education institutions from their educational goals. Similarly, three quarters of Chilean citizens believed that profit undermined the quality of education and were against colleges and universities being run for profit. These different definitions of education implied different roles for the state: those of a regulator of the education market, on one hand; or a guarantor and provider of education, on the other.

Free education and tax reform

Students argued that education should be free because it is a right. The government countered that support for free education amounted to having "*the rich subsidised by the poor*", since the tax monies required came from all citizens, rich and poor. By this argument, demonstrators were protecting the interests of students of elite universities, recipients of the biggest chunk of government funding, to the detriment of students from technical and professional schools, who self-financed their education and were, according to government speakers, working and studying rather than striking.

This argument led the student movement to focus on tax reform. The rich, they claimed, should support the education of the poor by paying higher taxes. The invocation of progressive tax reform, supported by nearly 80 per cent of public opinion (CERC 2011b), countered both the claim that free education was regressive and the argument that resources were insufficient. Indeed, students insisted that Chile had the resources to finance a broader education commitment as a result of the sustained high growth rates of the previous decades. From this perspective, the calamitous situation of education was the result not of poverty but of political decisions regarding income distribution in a quite prosperous land.

Principles and bargaining; intransigence and dialogue

The underlying differences in definitions and logics made negotiation difficult: while the government promised to reschedule debts and alleviate indebtedness, students called for a much deeper and more systemic reform. The fight was not, they insisted, for more resources but for a change of paradigm. The government saw student intransigence as "*ideological*"; they suggested that the students had "*lost their way*" and moved far beyond "*legitimate demands*" when raising issues such as tax or constitutional reform (Education Minister Joaquín Lavín, interview in *La Tercera*, July 2, 2011). In return, student leaders argued that their movement was indeed political because the required changes involved politics; they also focused their attention on the ideological assumptions hidden behind the government's proposals.

Nevertheless, over time government officials chose to minimise differences and repeatedly insisted that "*what brings us together is far more significant than what sets us apart*" (Education Minister Felipe Bulnes, in *La Tercera*, August 22, 2011). Clearly, however, dialogue and street pressure were understood as mutually exclusive by one side and as complementary by the other: while the government sought to deactivate street mobilisation by resorting to dialogue tables, students viewed street protest as an effective way to force their demands onto the dialogue agenda.

The past, the future, and fear

Students argued that they shared neither the statist ideas of the sixties and seventies nor the business ideology of the eighties and nineties. They insisted, however, that they did want to restore the free education system that had been enjoyed by the same political leaders who now wanted it privatised.

The opposition between the past and the future that eventually prevailed, however, was the one pitting dictatorship against democracy. The turning point occurred in August, when the government threatened to control demonstrations by using the State Security Law and giving free rein to police repression. The students characterised the government as "*heir of the dictatorship*" and their struggle, in contrast, as the heir to the struggles for democratisation (FECH Vice-president Francisco Figueroa, in *La Tercera*, August 26, 2011). Within that frame they called for reform of the Constitution, a Pinochet-era text that would have never been passed under a democratic regime. A key element in this development was the lack of fear by a generation born and raised in democracy; unlike their parents, indeed, the young were not afraid to denounce abuse and repression and, above all, did not fear that radical change could arouse a reaction amounting to a reversal to dictatorship.

The transversal strategy and the silent majority

The student movement's transversal strategy aimed at penetrating society from side to side with its universal claim to public education as an opportunity equaliser in an extremely unequal society, and its willingness to ally with other social actors similarly hurt by neoliberal policies. The students' demands potentially benefitted everybody; therefore, they argued, their movement was not just a student movement but, more exactly, a social and citizen movement.

Broad goals made it easier to include more social actors, which in turn granted legitimacy to increasingly ambitious demands. The result was a heterogeneous alliance of actors joined by an overarching cause. The alliance experienced rapid growth: it began with the universities (public and private; students, professors, deans, and workers); it expanded to the whole education sector (high schools, technical schools, parents, and guardians); it moved on to include workers' organisations; and eventually it became a "citizen movement". The so-called "social movement for public education" took shape by the end of June. The overwhelming support received from the public – a gigantic innovation in a fragmented and individualistic society – grew from the experience of common suffering in an unequal social system based on a highly unregulated market. Links with unions, in particular, were shaped by workers' concerns about their children's education as well as students' concerns regarding their own unpromising future as workers.

The government challenged the students' claim that the protest was society-wide and characterised it instead as representing just the "*5 per cent of most privileged students*" who "*make a loud noise*" (President Piñera, Interview with *Bloomberg Market*, in *La Tercera*, December 22, 2011). It also sought to exploit dissident voices among students and claimed to protect the "*right to study of the silent majorities*" (Santiago Mayor Pablo Zalaquett, in *La Tercera*, June 20, 2011). The silent majorities were later embodied by disorder-weary neighbours and were increasingly invoked to deny permits to march. As a result, the number of unauthorised marches grew – and so did police repression.

Cycles of violence and repression

Mobilisations often ended with violent clashes between masked *encapuchados* and police officers, and occasionally with looted stores and burnt vehicles. All protests, even peaceful ones, ended up with numerous arrests. Violence typically occurred in two contexts. Confrontations began when police intervened to disperse protesters because they "shouldn't be there" or to stop unauthorised marches. They also emerged as police evicted protesters from occupations.

Authorities justified evictions on the grounds that they were necessary to stop disorderly behaviour and damages; students claimed that most damage occurred during the evictions. In some cases they even reported police planting Molotov cocktails and provoking damage to incriminate students. Moreover, violence sometimes started simply because the police were there; their very presence sparked confrontations that would otherwise not have occurred.

On the other hand, *encapuchados* often took advantage of a police presence to initiate violence or provoke pitched battles. The people behind the masks were mostly students, although they were not responsive to student organisations. Observers typically described them as the casualties of an exclusionary society. Although they were just a small minority, the media tended to assimilate them to the whole student movement, negatively portraying hundreds of thousands of demonstrators with bare faces and no weapons.

Student leaders of all ideologies denied links with the *encapuchados* and emphasised their commitment to peaceful demonstrations. Demonstrators often attempted to stop violence themselves on the basis that "*we are not criminals*" and that criminalisation of the movement eroded public support. On the other hand, movement leaders also pointed out that lasting social order and peace could only result from justice. They denounced police officers that infiltrated marches to encourage violence, police failure to contain conflicts, and their tendency to apply repressive measures indiscriminately to both peaceful and violent demonstrators. Public opinion, in turn, attributed violence to police lack of expertise and government intransigence rather than to students' irresponsibility (Cooperativa, Imaginacción, and UTFSM 2011a).

Early on, police repression was denounced by the students as a deliberate government policy. From mid-June, reports were filed of excessive violence in marches and evictions, illegal arrests, assaults, beatings, threats, and abuses of authority. Repression and human rights violations were condemned by the National Institute of Human Rights and the Inter-American Commission of Human Rights (IACHR). Government officials, however, never stopped praising the police for their "professionalism" and "contribution to public order".

Criminalisation and the mass media

Despite evidence that the small groups of *encapuchados* did not respond to the student movement, the government insisted that the latter was responsible for all acts of violence, so social protest became a public order issue. Officials argued that marches, occupations, and strikes were forms of disorder as were stone throwing and Molotov bombs; official behaviour, in contrast, was portrayed as embodying civility, dialogue, and responsible work conducive to problem solving. The political opposition suggested that it was the criminalisation of social protest what prevented dialogue, as it did not allow for the acknowledgment of a "*profound demand of Chilean society*" (Senate President Guido Girardi, in *Radio Cooperativa*, August 18, 2011). But the official account prevailed in the media: news channels tended to illustrate all reports about the movement with a single photo symbolising violence – a practice that was denounced by the students.

The occupation of public space and the problem of order

Mobilisations reflected a struggle over the use of public space. Authorisations were required for assembling in public spaces in order to petition the authorities; therefore, police justified dispersing unauthorised demonstrations with force as "*according to regulations*" (*La Tercera*, June 8, 2011). Since secondary students were reluctant to follow the stipulated authorisation procedures, they were more often involved in unauthorised demonstrations and violent repression. Eventually the student movement began to question the need for authorisations under a true democracy. In July and August public officials began to forbid demonstrations; the student movement, backed by a number of opposition legislators, sued to protect their constitutional right to protest. The appeal, however, was dismissed by the Court of Appeals.

With or without authorisation, students insisted on marching. Officials accused them of acting above the law and stressed the government's responsibility to preserve order and peace. As the 9 August strike approached, the city administration accepted a marching route along Alameda Avenue, the most disputed public space. But the mayor insisted that marches scared neighbours and harmed the historic district. In October the police dispersed a demonstration of secondary students whose permit request had been submitted after the deadline. Two days later a CONFECH march was also repressed because the route had not been agreed upon. The issue catalysed debate within Congress, as opposition lawmakers called for replacing repression with concern for the causes of the conflict. By the end of the year, CONFECH leaders decided not to request authorisation for their Christmas protest because, they insisted, no permission was required to just walk on the city's sidewalks.

Distrust of politicians

Even the most moderate student leaders distrusted government and opposition politicians alike, due to their allegedly common preference for "*making arrangements between four walls*" (FECH Vice-president Francisco Figueroa, in *La Tercera*, August 8, 2011). This distrust was widely shared by the citizenry, with confidence levels of around 10 per cent for political parties and 17 per cent for Congress (CERC 2011a; 2011b).

Citizen distrust, however, was most often passively expressed in polls rather than through mobilisation. In that context, student leaders understood early enough the need for backing from opposition politicians, whose support included denouncing the government's lack of response to social demands, insisting on dialogues that included all social actors, rejecting criminalisation of social protest, and repudiating repressive measures. In this respect, opposition politicians and government officials did in fact display radically different conceptions of order, public space, and the legitimate use of force.

Political representation and social legitimacy

Opposition support for the student movement contributed to harsh controversies around political representation and legitimacy. While the student movement was sceptical about politicians and the potential for "dialogue tables" to legitimate decisions made behind closed doors, they welcomed opposition support in challenging government policies.

The debate about relations between political representation and social demands was unleashed when opposition leaders agreed to halt congressional debate of the education bills until the issues were discussed at dialogue tables. The government, in contrast, refused to yield to "*yells in the streets*" (Education Minister Joaquín Lavín, in *La Tercera*, July 2, 2011), as the issues were for Congress rather than the students to decide. The students' demands for dialogue conditions were then rejected as "*out of place*" (Minister Andrés Chadwick, in *La Tercera*, August 28, 2011). Opposition leaders, on the contrary, supported student demands and were accused of failing to take on their political responsibilities and "*hiding under the skirts of the student leaders*" (Senator Jovino Novoa, in *La Tercera*, October 10, 2011).

These debates reflected fundamental philosophical differences. On the government side, the emphasis was placed on the representative role of the government, authorised to make binding decisions by election results. On the movement side, the focus was on the excessive autonomy of politicians to legislate behind the backs of the people. From their perspective, legitimacy stemmed not just from vote counting but also from the validity of a majority claim for the enactment of a right.

The issue of representation and legitimacy also came up in congressional debates. The students increasingly recognised the importance of making Congress more representative, both by registration drives to increase youth electoral participation and by nominating their own candidates to run within existing parties – an option that sparked more dissent than support within the movement.

Conclusions

The outbreak of the student movement, its intensity, duration, and characteristics took both political parties and traditional civil society organisations by surprise. Few observers – or participants, for that matter – expected that it would last as long or have as much impact as it did. Also disconcerting for established political actors was its autonomy from political parties. Historically, political parties have mobilised the main social cleavages in Chile; therefore, the resulting social movements – such as that of the organised working class – have maintained strong, organic links with specific parties. This student movement, in contrast, was the first massive social movement not to be mobilised by a political party. The partisan affiliation of some of its more visible leaders should not be confused with leadership by that political party – a party that typically obtains a negligible proportion of the vote and is unable to summon more than a few hundreds of believers to the streets on any particular occasion. Not surprisingly, all attempts at using the students' popularity to advance partisan electoral interests failed miserably.

As a challenge to the political system, the students' protest came from the most unexpected place, as the majority of young people had hitherto followed the overall trend of rising consumerism and political apathy predominant after the dictatorship. In addition, in the process of posing their challenge, students made use of a variety of strategies drawn from an increasingly globalised repertoire that had the effect of making their protests look more alike those of their peers around the globe than those of their own parents three decades previously.

The deep causes of the conflict were structural and longstanding, so what prompted this eruption? And what made it different from earlier student mobilisations? Some interviewees point at collective learning processes from previous experiences that produced suspicion of government co-optation as well as willingness to get involved. Others emphasise the characteristics of the government, the first right-wing administration in 20 years, which tended to behave as "*businessmen … facing a union*", using "*pure logics of labour negotiation*" (Male university teacher and PhD student, interview).

Initiated in April and led by charismatic and telegenic youths, the movement peaked in August 2011. It utilised a large repertoire of actions and found legitimacy in the innovative use of colour, sound, movement, and imagination as well as its growing numbers of supporters. Authorities initially saw students as wayward children, characterised their marches as disorderly, and readily repressed them; however, their demands attracted widespread citizen support as responses to a real, serious problem. Indeed, the perception of education as the main national issue increased among all social strata from 24 per cent in May to 73 per cent in August 2011, according to CERC (2011a; 2011b). This perception translated both into a high turnout at demonstrations and a strong social consensus centred on a few core ideas and proposals. In that sense it can be argued that the student movement moved beyond pure protest in order to advance a few (admittedly underdeveloped) proposals – including that of a tax reform that would allow for funding to materialise in the quantities needed for a deep educational reform to take place.

A majority of citizens recognised that the students' concerns did not just belong to the students. After all, most families included present and/or future students. And the education model reflected and reinforced the elitist model of society bequeathed by dictatorship and maintained under democracy; therefore, education reform also implied political and social reform. Support for reform grew out of the frustrations accumulated during two decades. The expansion of demands – which incorporated the concerns of many additional constituencies – and the broadening of stakeholder participation occurred at a surprising speed. Many people had their first political experience alongside the students, and went on to state that this participation would not be their last.

As the conflict deepened, the Ministries of the Interior, Economy, Labour, Transport, Health, and even Foreign Affairs became gradually involved. The conflict involved not only the Executive but also the Legislative and Judicial branches of government, and both national and local authorities. Once they became impossible to ignore, the mainstream strategy was to thank students for putting the issue onto the agenda. However, the students resisted the view that once raised the issue should be solved by politicians. The failure of dialogue attempts generated reciprocal accusations of intransigence.

Participants, in turn, credited the movement with turning them from passive objects of public policy into political actors. Many saw the events as a historic, foundational moment that awakened citizens to questions about the basic principles of society. But it soon became apparent that starting from scratch was not that easy. It was unrealistic to remain on strike and maintain the occupations until the ultimate goals were achieved; alternative approaches were deemed necessary. Self-criticism identified several mistakes, including too much prominence of some leaders and too much emphasis on forms of mobilisation with high costs to students and low returns in social support. It was agreed that the movement should have formulated more concrete proposals prior to engaging Congress and been better prepared to compromise and negotiate agreements. Deepening the transversal strategy was therefore emphasised as a way of democratising demands at the grassroots level and strengthening joint work among various student groups.

In the meantime, Congress continued to discuss the bills introduced by the Executive and students resumed mobilisation in the hope of influencing that discussion. They sought improvements in the educational system while not losing sight of their long-term objectives; and tried to maintain the presence in the public space that had turned them into relevant political actors while trying not to lose the favour of public opinion. In March 2012 classes as usual were resumed, and mobilisation as usual was resumed as well. The movement, however, had lost momentum. It had managed to raise concern among the population and prompt a much-needed debate, but no major change had yet been obtained. As of today, it remains uncertain whether the ongoing process will in the long run produce more democracy and social change, or just more of the same.

Note

1. Media sources used in this article include *BBC Mundo* (www.bbc.co.uk/mundo), *BioBioChile* (www.biobiochile.cl), *Cambio 21* (www.cambio21.cl), *Diario de Antofagasta* (www.diarioantofagasta.cl), *El Ciudadano* (www.elciudadano.cl), *El Dínamo* (www.eldinamo.cl), *El Mercurio* (www.emol.com), *El Mostrador* (www.elmostrador.cl), *El Repuertero* (www.elrepuertero.cl), *La Nación* (www.lanacion.cl), *La Segunda* (www.lasegunda.com), *La Tercera* (www.latercera.com), *Publimetro* (www.publimetro.cl), *Pura Noticia* (www.puranoticia.cl), *Radio ADN* (www.adnradio.cl), *Radio Cooperativa* (www.cooperativa.cl), *Radio Tierra* (www.radiotierra.com), *Radio Universidad de Chile* (www.radio.uchile.cl), *The Clinic* (www.theclinic.cl), *The Guardian* (www.guardian.co.uk).

References

CEP. 2011. *Estudio Nacional de Opinión Pública Noviembre–Diciembre 2011*. http://www.cepchile.cl/dms/lang_1/doc_4936.html.

CERC. 2011a. *Barómetro de la política*, May–June. http://www.cerc.cl/pdf/Barometro_de_la_Politica.pdf.

CERC. 2011b. *Barómetro de la política*, August-September. http://www.cerc.cl/pdf/barometro_de_la_politica_agosto-septiembre2011.pdf.

Cooperativa, Imaginacción, and UTFSM. 2011a. *Encuesta Violencia en las manifestaciones y marchas*, October 24. http://www.imaginaccion.cl/encuestas2011/251011.pdf.

Cooperativa, Imaginacción, and UTFSM. 2011b. *Encuesta Evaluación año 2011*, December 12. http://www.imaginaccion.cl/encuestas2011/14122011.pdf.

UN. 2011. "Palabras de S.E. el Presidente de la Republica, Sebastian Piñera, con motivo del debate general en la 66° Asamblea General de las Naciones Unidas." [Words of the President of the Republic of Chile, Sebastian Piñera, the general debate on the occasion of the 66th General Assembly of the United Nations." http://gadebate.un.org/sites/default/files/gastatements/66/CL_es.pdf.

The author

Inés M. Pousadela is an Argentinean political scientist now residing in Washington, DC. A former professor at the University of Buenos Aires and visiting researcher at Georgetown University and the University of Maryland, she is currently a research fellow at American University and an independent political consultant for NGOs and international organizations. She has published books on political representation and participation in Argentina and Latin America, as well as a number of articles and book chapters on political theory, democratisation, social mobilisation, representation, political culture, and the issues of corruption, transparency, and accountability.

From embarrassing objects to subjects of rights: the Argentine LGBT movement and the Equal Marriage and Gender Identity laws

Inés M. Pousadela

This article examines the Argentine LGBT movement within a framework of two longer-term processes. First, an enlargement of rights that is typical of post-transitional situations, with the Argentine peculiarity that, a quarter of a century after its transition to democracy, the country experienced a strong resurgence of the human rights discourse while judicial procedures were reopened to deal with human rights violations committed under the 1976–83 dictatorship. Second, the global widening of the concept and practice of human rights and the fight against discrimination – including those related to sexual orientation and gender identity.

D'objets sources de gêne à sujets de droits: le mouvement LGBT argentin et les lois sur l'égalité en matière de mariage et d'identité de genre

Cet article examine le mouvement LGBT argentin dans un cadre englobant deux processus à plus long terme. En premier lieu, un élargissement des droits qui est typique des situations post-transitionnelles, avec cette particularité propre à l'Argentine qu'un quart de siècle après sa transition vers la démocratie, le pays a connu une forte reprise du discours des droits de l'homme tandis que des procédures judiciaires étaient rouvertes pour remédier aux violations des droits de l'homme commises sous la dictature de 1976 à 1983. En deuxième lieu, l'élargissement mondial du concept et de la pratique des droits de l'homme et de la lutte contre la discrimination – y compris les discriminations liées à l'orientation sexuelle et à l'identité de genre.

De objetos penosos a sujetos con derechos: el movimiento LGBT argentino y las leyes de matrimonios iguales y de identidad de género

El presente artículo examina al movimiento LGBT argentino en el marco de dos procesos de más largo plazo. En primer lugar, se trata de la restauración de derechos, típica de las situaciones postransicionales, con la peculiaridad argentina de que, 25 años después de su transición a la democracia, el país experimentó un fuerte resurgimiento del discurso de derechos humanos, a la vez que se volvieron a abrir procesos judiciales en torno a las violaciones a los derechos humanos cometidas durante la dictadura, entre 1976 y 1983. El segundo proceso tiene que ver con la ampliación del concepto y de la práctica de los derechos humanos, así como con la lucha contra la discriminación por motivos de orientación sexual y de identidad de género, entre otros.

De questões constrangedoras a questões de direitos: o movimento LGBT argentino e as Leis de Casamento Igualitário e Identidade de Gênero

Este artigo examina o movimento LGBT argentino dentro de uma estrutura de dois processos de mais longo prazo. Primeiro, um aumento de direitos que é típico de situações pós-transição, com a peculiaridade argentina de que, um quarto de século depois de sua transição para a democracia, o país vivenciou um forte ressurgimento do discurso de direitos humanos, enquanto procedimentos judiciais foram reabertos para lidar com as violações de direitos humanos cometidas durante a ditadura de 1976–83. Segundo, a ampliação global do conceito e prática de direitos humanos e a luta contra a discriminação – incluindo aquelas relativas à orientação sexual e identidade de gênero.

Introduction

This article examines the Argentine LGBT movement within the framework of two longer-term processes. First, an enlargement of rights that is typical of post-transitional situations, with the peculiarity that a quarter of a century after its transition to democracy Argentina experienced a strong resurgence of the human rights discourse as judicial procedures were reopened to deal with human rights violations committed under the 1976–83 dictatorship. Second, the global widening of the concept and practice of human rights – including those related to sexual orientation and gender identity – and the fight against all forms of discrimination.

Both processes occurred almost simultaneously in Argentina. When the dictatorship began in the mid-1970s, sexuality was not yet secured as a human right anywhere in the world. As democracy was restored in the early 1980s, sexual rights had just begun to be understood as rights, and the first Argentine sexual diversity organisation, the Argentine Homosexual Community (CHA), was then founded.

Through the 1980s and until the mid-1990s, the nascent LGBT movement was primarily concerned with issues such as homophobia, police violence, and the HIV/AIDS epidemic. By 1997 there were effective HIV/AIDS treatments covered by the national health system, and since the mid-1990s greater advances were made regarding civil and social rights. For about a decade the CHA was the main LGBT organisation, with other groups gaining more visibility during the 1990s. Lesbians eventually obtained wider public visibility during the process leading to the Equal Marriage Law, and trans(sexual) people did so during the campaign for the Gender Identity Bill.

The period covered by this study begins in 2005, with a series of controversies regarding prostitution in Buenos Aires, and ends in 2012, when the Gender Identity Law was passed – two years after the Civil Code was reformed in order to legalise same-sex marriage. The whole process, pushed forward by the Argentine LGBT Federation (FALGBT) and other organisations, was amazingly fast. Also surprising were the deep changes operated in public opinion and the consensus that formed around equality and non-discrimination regarding gender and sexuality. In less than a decade public debate evolved from a discussion of the legality of police edicts to the legalisation of same-sex marriage and the recognition of every person's right to request new identity papers rectifying her public identity, along with the obligation of health service providers to cover body adjustment treatments. Progress was made from a situation in which sexual diversity was embodied – and subject to ridicule and contempt – in the stereotype of the mannered homosexual and the "shameful presence" of trans people, to a situation in which diversity was given legal recognition and civil rights were granted equally, regardless of sexual orientation or gender identity, bringing along new opportunities for the effective enjoyment of other rights.

This article is based on primary and secondary sources, including the websites of magazines and newspapers (notably *La Nación* [*LN*], *Clarín* [*CL*], and *Página/12* [*P/12*]), radio broadcasts, and TV programmes; online profiles, publications, and other materials of the organisations and campaigns; and, last but not least, in-depth interviews with movement leaders, activists, participants, and qualified informants.[1] It first gives a brief overview of the sexual diversity movement in Argentina, after which it goes on to describe the processes leading to the Equal Marriage and Gender Identity Laws. It then discusses the actors and alliances, as well as their main strategies and actions, before examining the formation of a social consensus and the resulting cultural change, and providing some concluding remarks regarding the potential for the constitution of a new egalitarian common sense.

The sexual diversity movement in Argentina

The main precedent of the current Argentine LGBT movement was the leftist revolutionary Homosexual Liberation Front (FLH), founded in 1971. The FLH's immediate forebear was the Group Our World, established in 1967 by lower-class homosexuals, mostly trade union activists. Soon after, their convergence with a group of intellectuals produced the FLH. Throughout its short life, the FLH gravitated between leftist Peronism and Marxism and advocated for both to include the homosexual community's claims.

Relations between leftist and sexual diversity activism, however, were not without tensions. When the FLH staged its "*first heroic act of visibility*" (César Cigliutti in *CL*, 9 August 2010) by joining a demonstration in Plaza de Mayo, for instance, it was greeted by the chant "*we are not fags, we are not junkies, we are soldiers of FAR and Montoneros!*" In any case, what remained of the FLH dissolved as the dictatorship began, and with the end of the military regime the CHA emerged. "*Democracy began and we all thought that all guarantees and individual freedoms were automatically restored, but reality taught us that many things from the [dictatorship] remained*", explains Marcelo Suntheim. He went on to say that:

> "Murders of transvestite girls were quite frequent ... nothing was ever investigated ... It was enough to look gay while walking down Santa Fe Avenue for the police to stop [you] and apply the famous police edict ... that stated that scandal was not allowed in the streets." (Marcelo Suntheim, interview in *El Puente* 17, 2 September 2011)

The CHA initially focused on fighting police edicts and increasing the community's visibility. Since 1986, when they joined the 24 March event marking the tenth anniversary of the military coup, they placed the issue of sexual diversity within the framework of human rights. The Mothers' Resistance Marches, International Women's Day, and the March 24 mobilisations all became key landmark dates of the movement. Despite its intense activity, however, the CHA was officially recognised only in 1992, after years of judicial litigation.

The 1990s were marked by increased social visibility and intense work with the gay community. Within the latter women were still invisible, while "*even in gay settings, transvestites were just a part of the show*" (Mariano, interview 2 August 2012). This situation started to change, and the term transgender was coined in order to designate transvestite, transsexual, and intersexual persons. In 1992 the first Pride Parade (*Marcha del Orgullo*) was staged in Argentina, amid fears and a low turnout.

In 1993 the CHA ran a campaign aimed at forcing politicians to state their positions regarding sexual diversity, which has since been repeated before every election. In 1996 their anti-discrimination efforts were rewarded when Buenos Aires city's Constituent Assembly established sexual orientation as an unacceptable discriminatory excuse; two years later police edicts were abolished in the district. In 1999 the CHA sued the government so it would recognise the pension rights of same-sex partners. In 2000 a civil union bill was introduced in the local legislature, and two years later Buenos Aires was first in the region to have a law granting legal recognition to same-sex couples.

In the 1990s sexual diversity organisations began to multiply and spread beyond the city of Buenos Aires. A heterogeneous movement was created that included specific organisations for male gays, trans people, women and transvestites involved in prostitution, feminist lesbians and members of co-maternal families, as well as the novel "diversity branches" within most political parties. This organisational process took place within a wider global process, the first milestone of which occurred in 1990, when the World Health Organization removed homosexuality from its list of mental illnesses. A decade later, in 2001, the Netherlands became the first country in the world to legalise same-sex marriage, and others soon followed. Inspiration came to the Argentine LGBT movement when a similar law was enacted in Spain, in 2005.

Annoying objects

As of 2005, transvestites were regarded as men who dressed up as women in order to prostitute themselves. Only seven years later, when the Gender Identity Law was passed, they had begun to be seen – to a great extent due to the FALGBT's work – as persons whose gender identification did not fit the socially sanctioned identity they had been assigned at birth, and who were mostly involved in prostitution due to lack of alternative options.

In 1996 the new Constitution of the city of Buenos Aires had recognised the "right to be different" and included gender and sexual orientation in its antidiscrimination clause. In January 2005, a new Code of Misdemeanors authorised the offer of sex in public places located at least 200 metres away from private residences and educational and religious institutions, leading transvestites to move their activity towards the Rosedal, a big park in the Palermo quarter (*CL*, 27 January 2005). Their presence triggered lengthy controversies that led to prohibition and protest.

Following complaints from a group of neighbours, the city government banned the offer of sex in the Rosedal in July 2007. This decision was temporarily suspended after a demonstration took place outside the city government offices, and the organisations involved were summoned to a "dialogue table". Here, organisations representing sexual diversity demanded that the local government implement health, education, and employment policies for trans people, whose life expectancy was calculated at about 35 years (*CL*, 4 August 2007). In mid-2008 an agreement was reached to relocate prostitution to a more isolated area of the park. The agreement included new toilets, trashcans, and speed bumps; a health programme; a schedule for sex trade; and reinforced police presence (*P/12*, 7 May 2008). Annoyed neighbours were now a minority: many more emphasised the security gains brought by the presence of "the girls".

In the meantime, the demand by ATTTA (Association of Argentinean Transvestites, Transsexuals, and Transgenders) that would eventually translate into the Gender Identity Law gained increased public visibility. In September 2008 a decision by a Mar del Plata judge authorised Tania Luna to get a new female ID without undergoing surgery. Immediately after, ATTTA's president Marcela Romero obtained a new ID after a ten-year judicial process.

In 2009, a Gender Identity bill was introduced. According to a survey conducted by Analogías in 2008, 66.6 per cent of the inhabitants of the country's main cities were in favour of trans people being allowed to change the name on their identity papers, with only 22.5 per cent against. Even more categorically, 75.7 per cent agreed that sex reassignment surgeries should be allowed (*P/12*, 1 April 2010). A process thus began that culminated in May 2012 with the recognition of the so-called "right to be".

The same rights with the same names

In June 2005 homosexual marriage was legalised in Spain, a Catholic country seen as culturally conservative and with strong historical and cultural ties with Argentina – but led by a socialist government. That same year the Argentine Congress ignored a CHA initiative regarding the extension of civil union to the national level. Soon after that, in June 2006, the FALGBT was founded, and three months later representative Eduardo Di Pollina drafted a bill aimed at reforming the national Civil Code to allow marriage to all couples, regardless of sexual orientation or gender identity. The bill was introduced in April 2007 with the signatures of representatives from most parties, and a similar one was introduced in the Senate soon after. In November the 16th edition of the Pride Parade took place under the motto "*Our celebration is a demand. Equality, Freedom, Diversity*" (*P/12*, 16 November 2007).

Earlier that year, FALGBT president María Rachid and her partner had gone to the Registry Office in Buenos Aires to request an appointment to get married (*P/12*, 15 February 2007). As expected, they were rejected, so the FALGBT interposed a *recurso de amparo* demanding compliance with the constitutional provision of equality. From then on, the judicial route was used to complement the congressional strategy. In February 2008, the Rashid and Castro presentation reached the Supreme Court, garnering much media attention; soon after, a Congressional Front for the Rights of Sexual Diversity was formed by a group of legislators from all parties.

Since the presentation of the first *amparo* a social consensus was discernible, at least in Buenos Aires and its metro area. According to a poll by Analogías, 73 per cent of the citizenry were in favour of same-sex marriage; even more people (82 per cent) favoured civil union. "*Society is very much ahead of institutions*," pointed out María Rachid.

> "We often hear legislators tell us that they agree with us … but that this is not the right time, as if our lives could be put on hold until they believe that the moment has come to stop discriminating against us. Clearly, the people are ready, and what's needed is for their representatives to take note of the message that this survey is sending them."
> (María Rachid, interview in *P/12*, 18 February 2007)

The bills were never debated and were reintroduced. Continued efforts to engage both the courts and legislators helped move the debate forwards. And in November 2009 tens of thousands of people took part in the 18th edition of the Pride Parade (*CL*, 7 November 2009). The first milestone took place in the judicial front just a few days later: in response to the Freyre and Di Bello *amparo*, judge Gabriela Seijas ordered the Buenos Aires Registry Office to marry the couple, declaring the Civil Code articles that defined marriage as the union between a man and a woman to be unconstitutional. Catholic lawyers' entities and other denominational organisations criticised the decision and lodged several, unsuccessful, appeals while Freyre and Di Bello obtained an appointment to get married on 1 December. Before that date, however, Seijas' ruling was declared void by a judge from a different jurisdiction. Freyre and Di Bello eventually managed to get married in Tierra del Fuego, their success strengthening the legal demands of many others.

Meanwhile, the FALGBT showcased their political support with a press conference in the Congress involving representatives from five different political parties. Soon afterwards a joint meeting was summoned of the two congressional committees involved, and the resulting bill was sent to the House, where it underwent additional delays. Finally, on 5 May the House passed the bill with 126 votes for and 110 against; hundreds of demonstrators celebrated in the streets.

A few weeks later, the first in a series of public hearings on the bill took place at the Senate's General Legislation Committee, with interventions from representatives of several organisations on both sides. Besides arguing that homosexuality was against nature and that the etymology of marriage excluded the possibility of two people of the same sex entering the institution, opponents to the reform of the Civil Code insisted that same-sex marriage was foreign to the Argentine cultural roots and was instead part of the so-called *agenda of the city of Buenos Aires*, more subject to foreign fads and influences than the country's healthier interior provinces. The Catholic Church's Commission of Legislative Monitoring successfully requested that public hearings be held in several such provinces. As a result, the Committee toured a dozen provinces, where its very Catholic president held meetings with religious leaders while systematically refusing to receive LGBT leaders.

LGBT voices were downplayed in the hearings, while those of Catholic and Evangelical representatives were disproportionately heard. In response, the FALGBT organised a series of activities that included demonstrations, cultural events, and meetings with leaders of other social organisations, and concluded with a festival on 28 June, International Gay Pride Day, which summoned 50,000 people. Meanwhile, the judicial strategy kept on yielding: thanks to a favourable ruling by a La Plata judge, in late June a wedding between two women took place in the province of Buenos Aires.

After eight hearings in the Senate and a dozen in several provinces, the Committee produced a report recommending the Senate decide whether to discuss the bill passed by the House, whilst offering an alternative civil union project. The bill then received a decisive boost: former president (then national legislator) Néstor Kirchner stated his position in favour of legal change and exhorted the nation to "*leave all discriminatory and obscurantist visions behind*" (*CL*, 8 July 2010). President Cristina Kirchner intentionally missed the traditional Independence Day religious service after a negative reaction from the Catholic Church's top authority, and instead gave a speech in which she demanded that an egalitarian society be built in which "*minorities have the same rights as the majority*" (*CL*, 9 July 2012; 12 July 2010).

On 14 July, the Senate passed the bill without modifications and Argentina became the first Latin American country to legalise gay marriage, unleashing euphoric celebrations. "*Argentina is now a better country*", announced Alejandro Freyre. The change was, he emphasised, "*not just for gays and lesbians, but for everybody*" to celebrate (*CL*, 15 July 2010). The president signed the bill in a celebratory event that was attended by LGBT activists, human rights leaders, judges, ministers, and governors.

By the end of the year same-sex marriages had taken place in all of the provinces; in every district, the first couple to marry made the news. By the law's first anniversary there were 2,697 same-sex married couples (60 per cent male, 40 per cent female) (*LN*, 14 July 2011). "*Now that TV cameras do not go to the Registry Offices anymore, a lot more people gather the courage to request an appointment*," assured Paulón (*CL*, 30 January 2011). The increased visibility of the LGBT community in places that were resistant to diversity was celebrated as an encouragement for others to "come out of the closet".

The right to be

As soon as the House passed the equal marriage law, the activists started chanting: "*Legal papers for transvestites and transsexuals!*" (*P/12*, 6 May 2010). A campaign was launched for a law of Gender Identity and Health Coverage for trans people right after the Senate vote. "*Consensus is there even before we start looking for it*", remarked trans activist Alba Rueda (*P/12*, 22 July 2010).

The first Gender Identity bill (2007) had expired without debate and had been reintroduced in 2009. Discussion of this bill started in the House of Representatives, and ATTTA activists were for the first time invited to express their demands. The FALGBT used a similar legal strategy to that which had helped speed up congressional discussion and approval of the Equal Marriage Law. New IDs for trans people were requested through the courts. And in November 2010 the National Front for the Gender Identity Law, recently formed by several organisations, had their own initiative introduced in the National Congress with the support of representatives from several parties.

That year, the 19th Pride Parade's motto was "*Let's go for more. Gender Identity Law NOW.*" The ambience was more relaxed than in the past – and more political too. As pointed out by a journalist, "*it felt like that there was no need to put up a show, as could be seen in the case of many transvestites who turned up wearing their everyday women's clothes*" (*CL*, 7 November 2010).

In December a favourable response was given to a court request filed on behalf of trans actress Florencia de la V. The judge ruled that identity was a human right and dismissed the imposition of any psychiatric or medical requirement to justify the demand. Later that month, another judge ordered the state to provide a person with new papers with the gender identity of her choice without proof of any surgery or medical procedure. "*A person's sexual identity goes far beyond the biological*", the judge stated, urging Congress to pass a law simplifying the procedure (*P/12*, 4 January 2011).

Committee discussion started in mid-August 2011. The debate in the House, however, was postponed due to upcoming presidential elections. Meanwhile, in late August the FALGBT presented its Plan for LGBT Citizenship along with the report *Legislation for Equality*, based on a survey on the LGBT situation in 27 Argentine cities. Immediately after that, the Ministry of Health launched a pilot programme that ran at night-time in public hospitals and provided medical check-ups and treatment to people – such as transvestites and transsexuals – who usually avoided such places during regular business hours.

In October new expressions of support started growing out of universities. The National University of Córdoba's Higher Council approved a proposal for the institution to respect the "self-perceived gender" of its students, teachers, and personnel, which allowed them to request that the name of their choice be used in all of the university's areas, including its hospitals. The lead was followed by other public universities. In November 2011 the Buenos Aires city Ombudsman Office mandated that their name of choice be used in all legal procedures involving trans persons, and required local and federal police forces to do the same. Shortly after, the Public Security Minister required all police and security forces to respect their members' gender self-perception and provide uniforms and restrooms accordingly. A year later, the Health Ministry of the province of Buenos Aires obliged all its public hospitals to call trans patients by their chosen name, which was also to be used in their clinical histories. This measure, which also included awareness training, was expected to have a deep impact, as 40 per cent of trans people had never visited a medical facility for shame or fear (*LN*, 116 December 2011; *P/12*, 17 December 2011).

Tens of thousands of people took part in the 2011 Pride Parade, and for the first time in Argentina the transvestites headed the parade, exhibiting placards of their new IDs and the phrase "my identity is a right". Three days later, in a room full of activists and television cameras, a plenary meeting of the two involved congressional committees took place; the resulting majority report was a synthesis of the many initiatives under consideration. Around the same time the National Institute of Statistics and Censuses (INDEC) launched its first national study of the trans population.

Thanks to an agreement among the main parties, the House passed the bill by 167 votes to 17 in the last session of 2011. A few months later, in May 2012, the Senate approved by unanimity a Gender Identity Law that was described as one of the most advanced in the world, for being based on the principles of de-judicialisation, de-pathologisation, decriminalisation, and destigmatisation of trans identities. The new law enshrined the state's obligation to provide every citizen with identity papers consistent with their self-perceived identity, regardless of biological sex and with no need for medical or psychiatric diagnoses or adaptation surgeries. Both adults and minors (with their parents' or judicial tutor's approval) were granted access to hormonal treatment with no need for judicial or administrative authorisations; treatment and surgeries would be provided through both the public system and private health insurance companies as part of the Compulsory Medical Plan.

Actors

The demands and initiatives regarding the human rights of LGBT persons that eventually translated into the Gender Identity and Equal Marriage laws had their origins within civil society, were pushed forward by a social movement in conjunction with a series of allies within the political system, and were embraced by a majority within the latter only when it became evident that a solid social consensus had formed around it. The leading role was played by the FALGBT, a national network that grew to about 60 organisations nationwide.

The most visible face of the demand for equal marriage was then-FALGBT president María Rachid. In December 2010 the Spanish newspaper *El País* named her among 100 prominent people from Spanish America, and she went on to be elected as a Buenos Aires legislator (on the party ticket of Kirchner's Frente para la Victoria – FpV). The Gender Identity bill, in turn, was driven by an alliance of organisations and granted unprecedented visibility to trans leaders such as Lohana Berkins (ALITT) and Marcela Romero (ATTTA).

The ability to build and maintain a unified front for the advancement of its demands was one of the movement's main assets – one not present in many other countries in the region. The network of organisations aligned behind the demand, in turn, grew as their efforts were met with conservative resistance.

> "The public hearings in the provinces caused a reaction ... People got organised in order to do something ... It also helped to tie nationwide alliances with unions, students', workers' and human rights organisations that never before had had any contact with a sexual diversity organisation, but who saw the demonstrations and listened to the atrocities ... and got angry." (María Rachid, interview in *P/12*, 18 July 2010)

The social movement's main allies within the institutional system were some judges (particularly female ones) who issued rulings ordering the Registry Office to marry applicants, as well as a few legislators (again, especially women legislators) who pushed forward the bills in both houses of Congress. Within the Executive branch the National Institute against Discrimination and Xenophobia (INADI) stood out; also significant was the support expressed by former and current presidents Néstor and Cristina Kirchner. Last but not least, the movement's allies included journalists, artists, students, academics, psychoanalysts, and even religious and trade union leaders. Also important at key moments was the presence of foreign activists such as Pedro Zerolo, who provided invaluable know-how from the Spanish experience.

On the opposite side were the hierarchies of the Catholic Church and a variety of evangelical churches, alongside related institutions, such as lawyers' organisations, professional bodies, schools, and universities. Most experts who argued against egalitarian marriage before the congressional committees came from these institutions.

Women in action

Throughout the process, women played a major part. The main driving forces in the National Congress were almost invariably women. Twelve out of 20 representatives who signed the Equal Marriage bill in early 2007 were women (*P/12*, 2 May 2007), as were both chairs of the legislative committees and the judges who issued the first rulings stating that the prohibition of same-sex marriage was unconstitutional. The provincial governor who facilitated the first same-sex marriage was a woman. Between 2006 and 2009 the INADI was led by a woman, the main visible face of the FALGBT was a woman, and a female couple was the first to request an appointment to get married.

Despite being underrepresented in all sites of institutional and social power, women stood out not only among the leading actors of the movement but also among its allies in positions of authority.

> "If we hadn't had gender quotas, the law wouldn't have passed. There were a hundred women legislators in the House of Representatives: two out of three voted in favour [of equal marriage]." (Ignacio, interview 1 August 2012)

To explain this support, our interviewees emphasise the parallels between the present situation of homosexuals and that historically endured by women, as well as the work still needed in both cases to turn legal equality into real social equality.

> "Thousands of years with lessened rights give women a different perspective ... Women were [our] great natural allies ... Heterosexual men rule the planet since the beginning of time, so they don't see this as urgent." (Ignacio, interview 1 August 2012)

> "There is a shared agenda, which we had even before the dictatorship. The feminist movement worked together with the sexual diversity movement ... Our lesbian comrades have historically been relevant participants of feminist groups ... With the advent of democracy ... they again took each other's agenda. Because [it is] the patriarchal and sexist idiosyncrasy [that] establishes inequality. Therefore, your agenda is my agenda." (José Maria Di Bello, interview 4 August 2012)

Far from having taken place "naturally", however, this shared agenda is the result of hard political work, and relations were not always smooth:

> "When I started speaking about transvestites within the women's movement, I was called all kinds of names. They said that I was coming with these men disguised as women in order to try and colonise a space that belonged to women." (Maria Rachid, interview in *Zona de Respuesta*, 12 March 2012)

Youth and generational change

Another element repeatedly empathised is the youth of the movement that propelled the legal change: both the leading role played by recently created organisations and networks, and the young age of the great majority of the LGBT activists who were mobilised during the process.

The age differential is apparent not just in the intensive use of new communication technologies but also in the growing distance between young activists and the "old homosexuals". The latter seem to be increasingly viewed by the new generation with a mix of reverence – given their role as pioneers – and impatience towards their inflexible, prejudiced stances. In contrast, the younger members of the Federation tend to see themselves as more pragmatic, change-friendly, stereotype-free, and as the carriers of a wider agenda. Hence their celebration of the citizenship rights entailed by widened access to the institution of marriage – the same institution that their elders (mostly from within the CHA) kept rejecting as conservative and bourgeois.

Political parties and leaders

"*Bills such as this one correspond to very personal stances and do not reflect the positions of party blocks*", declared senator Vilma Ibarra when introducing a bill on same-sex marriage (*P/12*, 16 October 2007). The main congressional driving forces of the reform of the Civil Code were a handful of leftist and centre-left legislators – only a few of whom belonged to the powerful FpV. Most of their political parties were small, so the final decision whether to allow the initiatives to be discussed fell to the majority party's leader – former president Néstor Kirchner. Our interviewees' views differ widely regarding Kirchner's actions and the principled or opportunistic character of his involvement; however, they all stress the importance of the role he played.

Within the framework of the Argentine political system the commitment of a party leader does not entail the automatic alignment of his party's legislators – especially when it is assumed that moral convictions are at stake. This means that votes for the marriage bill had to be secured one at a time – a task the Federation activists obsessively devoted to themselves (Bimbi 2010). On account of internal differences, all congressional parties granted a free vote to their members.

> "47 out of 87 Peronists voted in favour, as did 16 out of 42 Radicals, 5 out of 10 PRO representatives, and 7 out of 21 from the Peronismo Federal ... Those on the centre-left always did the right thing ... On the centre-right, the most liberal ones also supported the bill." (Esteban Paulón, interview 31 Jusly 2012)

Religion and churches: influences and dissidences

The opposition to the new marriage law was mostly religiously inspired. Although churches were mostly against it, their opposition was not unanimous: dissent was present in all of them, and there were even a few small ones on the reformist side. Thus the debate highlighted religious pluralism within Christianity and Catholicism.

As pointed out by Jones (2010), the world of evangelical churches in Argentina comprises both a liberationist pole with an ecumenical tradition and a history of human rights advocacy, and a biblical conservative pole accounting for the highest number of parishioners. The situation is different in the Catholic Church, due to its unified and pyramidal structure. In this case, differences became apparent between the stances taken by the Church's hierarchy – which, although not monolithic, were hostile to same-sex marriage – and those of some minority catholic sectors that were in favour. There were also so-called diversity groups within most if not all religions, such as Gay Christians or Gay Argentine Jews (JAG).

According to many LGBT activists, the stance adopted by the leaders of the Catholic Church ignored both internal dissent and the opinions of the majority of its parishioners. As shown in a poll by the Center for Studies and Research on Labor (CEIL), the opinions of the average Catholic (about 70 per cent of Argentineans declare to be Catholic, but most do not regularly attend religious service) on topics such as abortion, sex education in schools, the use of condoms, premarital sex, and public funding of denominational schools differ widely from the official positions of the Catholic Church (*P/12*, 18 July 2011).

Religion's political and social influence also varies markedly with geography: it is lowest in the city of Buenos Aires, its metropolitan area and a handful of other big cities – which, as a whole, account for about half the country's population. It is not surprising, therefore, that about 80 per cent of the votes in favour of equal marriage in the House came from the city and province of Buenos Aires and the province of Santa Fe. Despite the fact that the smallest and most conservative provinces are overrepresented in Congress, a process of secularisation and displacement of religious beliefs to the private sphere seems to be well underway, as expressed in the following statement by a legislator: "*I have my religious convictions, but today as a national representative I have to make a decision consistent with my democratic convictions*" (*P/12*, 16 April 2010).

Actions and strategies

The movement's public visibility increased exponentially with the deployment of two simultaneous and mutually reinforcing strategies, judicial and congressional.

The judicial strategy was launched in February 2008, and consisted of filing *amparos* on behalf of same-sex couples that had been denied a marriage appointment at the Registry Office. The complaint was based on the argument that the Civil Code articles that prevented them from getting married were unconstitutional. As a result of these, a number of judges ruled on the unconstitutionality of such discrimination. By July 2010, when the Equal Marriage Law was passed, nine marriages had taken place via court orders, despite the legal campaign orchestrated by lawyers from Catholic institutions.

Every new favourable ruling was brandished by the FALGBT as a call to attention for Congress. By mid-2010 the Supreme Court awaited a prompt congressional solution to the issue; in turn, the expectation of an imminent Supreme Court ruling functioned as an incentive for the federal government to push the issue forward and have it debated in Congress.

The congressional strategy started yielding in 2009, when debate on the marriage bill first took place. Ahead was a more difficult task than that of obtaining a favourable ruling from a progressive judge – that of aggregating the wills of dozens of legislators. The majority of lawmakers had no clear idea or position on the topic; they were, however, subject to hard-to-resist pressures based on the exploitation of their fear either of God or/and their own constituencies, presumably opposed to policies that were defined as antireligious, contrary to natural law, and destructive of the social fabric. Therefore, a cultural battle took place that was aimed at increasing visibility, disseminating reliable and accurate information, and fighting irrational fears. Its two simultaneous targets were the citizenry as a whole and their congressional representatives – who were in turn likely to change their minds if a favourable consensus became apparent among their voters.

These strategies were reinforced by parallel steps forward that resulted either from denunciations of discrimination filed by LGBT organisations and the INADI or from autonomous decisions adopted by government departments and educational institutions. These were supplemented with actions of high symbolic value, such as the awarding of tokens of public appreciation ("woman of the year", "outstanding citizen") to prominent LGBT activists.

The repertoire of actions

Especially relevant among the actions targeted to legislators were requests for individual meetings to explain the movement's demands to the reticent ones, and coordinate the work ahead with the gay-friendlier ones. Other actions aimed at (or organised with) legislators included open letters requiring them to give priority treatment to the bills, and press conferences held in the Congress.

Information on the issue of marriage and diverse families was compiled in publications that were endorsed by well-known researchers and psychiatrists, psychologists and health professionals and organisations, and presented in public events. The FALGBT also made a series of spots designed to show that celebrities, role models, and authorities within art, culture, science, and religion supported the laws. Their aim was to display an existing social consensus while simultaneously creating and/or consolidating it. In the mass media, the mostly favourable journalistic coverage was supplemented by a steady stream of contributions by political leaders, social activists, respected academics, and specialised professionals, typically in opinion columns and interviews. However, it was the presence of the movement's main figures on news programmes and televised debates that facilitated the wider diffusion of their demands and knowledge of their arguments amongst the general population. The court-ordered marriages were also staged as political events, including press conferences and starring LGBT activists surrounded by their organisations' leaders and lawyers, INADI officials, and national and local legislators.

In the streets, activities took place either on symbolic dates or at key moments in the political process. Amongst the former was the yearly Pride Parade, which summoned ever-increasing crowds and was turned into a political celebration. In 2010 and 2011 the parade included the unusual presence of a variety of party organisations such as Kirchner's La Cámpora or the leftist Partido Obrero; diversity groups newly created within political parties, such as Putos Peronistas ("Peronist Faggots", a good example of the strategy of appropriating an insult); and even union confederations. The International Gay Pride Day was also celebrated every year. In 2010 the event took place just two weeks before the decisive congressional session and was shaped as a demonstration ("March for Equality") crowned with a festival in the square adjacent to the National Congress. Other demonstrations were staged around the key congressional sessions; all of them were typically organised through social networks.

Chains of equivalence

One successful strategy was to forge alliances based on chains of equivalence among claims of rights. That became especially apparent in the 2010 celebration of International Gay Pride Day. The demonstration sought to show that the demand for equal marriage did not belong to a small group but to society as a whole. It was fairly successful, as about 50,000 people attended and it was endorsed by a number of social, trade union, student, and human rights organisations, as well as prominent artists and scientists (*LN*, 29 June 2010). Equivalences among claims for the full enjoyment of rights were brought up by LGBT leaders on numerous occasions:

> "This law can open the doors to others regarding migrants, handicapped people, or even regarding the issue of abortion decriminalisation, which in Argentina shares the same obstacles with same-sex marriage, since the Church wants to prevail over the rest of society." (María Rachid in *P/12*, 16 July 2010; cf. also *P/12*, 3 September 2010)

Similarly, during the first Pride Parade following the Civil Code reform, the slogan "*one struggle, all the struggles*" worked as an umbrella for a wide spectrum of civil society organisations. Having defeated one inequality, other inequalities moved towards centre stage, including ones seemingly unrelated to the LGBT cause:

> "We also have to fight against other inequalities related to poverty and the condition of women ... Discriminations pile up on top of each other: you can be discriminated

> against for being a lesbian, a woman, for being poor, for coming from an interior province. It is against that that we have to fight. There are much deeper struggles ahead." (Ignacio, interview 1 August 2012)

Demonstration effect and regional networks

Just as it learnt from the Spanish process, the Argentine LGBT movement in turn sought to create a regional "demonstration effect" and transfer its experience to other countries where debates on the topic were only beginning, such as Brazil, Chile, Colombia, and Uruguay. As stated by Brazilian congressional representative Jean Wyllys, the first openly gay legislator in his country, "*after passing this law, Argentina became a reference for the whole of Latin America*" (*P/12*, 15 July 2011).

The Argentine process had a multifaceted dynamic effect at the regional level. This effect was visible in the work of the transnational activism that operated in the social networks; the growing opportunities for exchange within the framework of regional forums such as the LGBT Network of Mercosur; and the participation of Argentine (as well as Spanish) activists and legislators in events designed to drive similar processes in neighbouring countries – such as the 2011 launch of the Brazilian Congressional Front for LGBT Citizenship. In addition, a mechanism started to operate at a regional level, enabled by international equal marriage – i.e. same-sex couples of foreign activists getting married in Argentina and subsequently demanding from the state in their own countries that their marriage licenses be recognised, therefore setting off legal battles potentially leading to same-sex marriage legalisation there. The first to use this mechanism were two Paraguayan activists, with the mayor of Rosario and former legislator Silvia Augsburger acting as their witnesses.

Activists, collaborators, and sympathisers

As emphasised by several interviewees, the process leading to the Equal Marriage and Gender Identity laws was highly labour-intensive, but had a minimal financial cost. The FALGBT president reports:

> "The equal marriage campaign cost nothing. 30,000 pesos for an event, and it was all donations … All artists played for free … We got free publicity … The filming [was] done out of the good will of a few friends who are artists' managers." (Esteban Paulón, interview 31 July 2012)

The Federation is simply a network of organisations. It has very limited resources of its own, most of which come from international cooperation institutions and are earmarked for specific projects. The Federation does not have its own office. During calm times its everyday operations require just a few hours of work by its president and the occasional activist (none of whom are paid). In contrast, during the gender identity and equal marriage campaigns, it required the full-time dedication of a number of activists – many of them public employees or on the payroll of political parties and social organisations. "*We do this with great commitment; this has an enormous personal cost*", underlined Paulón.

The campaigns were therefore based on the voluntary work of hundreds of people; only some of these think of themselves as activists, while many more describe themselves as collaborators or sympathisers. Among the latter were well-known faces and anonymous people, homosexuals as well as heterosexuals.

> "Our family members, our friends, all the people from the [party] bloc, they were all for the bill and did whatever they could so it was passed. There were lots of people involved. If it had been just us, it would have failed." (Ignacio, interview 1 August 2012)

> "There were heterosexual people who went to the demonstrations . . . and smaller events . . . People I knew from the University, who were there to show their solidarity . . . Many 'independent citizens' felt the appeal of this cause." (Santiago, interview 6 August 2012)

Cultural change and social consensus

The day before the same-sex marriage law was passed, María Rachid told the press that it was "*a won cultural battle, in which there is no way back*". She added: "*When we set out we did not imagine that we were going to have the consensus that we achieved*" (*P/12* 13 July 2010). When the process began, LGBT activists believed that the cultural battle within society would be long; they were therefore surprised when in 2008 the first survey on the topic produced a 63 per cent approval rate for their demand. From then on, "*expressions of support started arising that we had never imagined that we would get*" (María Rachid, cited in *P/12*, 18 July 2010).

According to the initial diagnosis:

> "among those against equal marriage were two groups: the majority of those who rejected it did so due to prejudice, disinformation, ignorance. And another, smaller group, that had an ideological motive, which means that they were defending their privileges." (María Rachid, cited in *P/12*, 2 January 2011)

This distinction had to be kept in mind in order to avoid "*[treating] as an enemy someone who is actually misinformed, as a result of their education, their background*" (María Rachid, cited in *P/12*, 2 January 2011). The campaign was therefore aimed at convincing this social majority by the force of the better argument and to fight their fears, typically based on stereotypes and biased (or false) information. The task, however, was not to be limited to the distribution of information; it involved "putting ourselves on the line":

> "We threw ourselves body and soul into the task of building social consensus . . . [After the first *amparos*] almost one and a half years passed and no other couple volunteered. Nobody dared do it because it meant a lot of public exposure . . . When December 1st came and we almost got married . . . there was already a social thing that had been built around interviews and media appearances . . . People said: '. . . let them get married'. We even met grandmothers who told us that they had *prayed* so that we would be allowed to marry! . . . Visibility was essential; we had to show who we are: two people who love each other, with all our defects and virtues." (José María Di Bello, interview 4 August 2012)

Lawmakers followed a similar trajectory to society in general, with a number of representatives confessing to having become more aware and having changed their views along the way. Change could be observed even among those who ended up voting against equal marriage: as Vilma Ibarra pointed out, "*the most troglodyte legislators defended the alternative of civil union as if excusing themselves for not voting in favor [of marriage], ashamed of what they were doing*" (*P/12*, 6 May 2010).

The media also underwent a learning process regarding diversity, and new concepts started to turn into common knowledge, including the distinction between sex and gender, the difference between transvestites and transsexuals, the relationship (and not just simple identification) between the members of this community and prostitution, and an understanding of the limited opportunities LGBT people enjoyed and the everyday difficulties that they faced. This progressed to such an extent that the media – and especially journalists – ended up being recognised by LGBT activists as invaluable allies.

The societal debate

"*There was no [TV] programme, no dinner table, no school where no debate took place*", remembers Paulón. This intense social debate put in the limelight a number of well-prepared activists with an articulate, consistent discourse that was also conciliatory and moderate.

> "We prepared ourselves ... We looked at Spain; we looked at how the law had been achieved in different countries, which arguments had been used, which arguments [against], how to reply to them ... During the first debates on television [with] priests, we got angry and ended up screaming ... In later debates we were absolutely calm ... we looked at that bitch that said all kinds of atrocities and answered: [in a sweet, low voice] 'Well, we think that the country has to move towards equality, that we all should have the same rights...' It was the right strategy because we won the social debate with that. The Federation spoke of equality, rights, inclusion, family, diversity, joy, happiness, and love."
>
> "We had a number of resources readily available. So if the Catholic Church said that we would all go to hell, we immediately brought to the Senate all the evangelical, Catholic, and Jewish leaders we had and held ... an intercultural mass for the equal marriage law. If it was said that the project belonged to the Kirchner's ... we organised a press conference with the [presidents of all main congressional parties] ... Then it was said that gays were not fit to raise kids: there we went with psychologists and psychiatrists, with their professional associations." (Esteban Paulón, interview 31 July 2012)
>
> "If we had gone more radical, if we hadn't had a discourse fit for the TV screen –more accommodating and pro-family, less anti-church-, this would have been a failure. We never stopped repeating that some of us were Catholics." (Ignacio, interview 1 August 2012)

Televised debates systematically exposed the argumentative asymmetry between the contenders. "*They prepared themselves very badly, we were very lucky*", admits our interviewee. Not many were willing to be seen as the standard-bearers of the denial of rights; on the contrary, "*the majority wanted to remain invisible ... Only a reactionary minority ... was willing to appear on television ... Only the most unpresentable ones came out*" (Ignacio, interview 1 August 2012).

The congressional debate replicated, only more systematically, the arguments already present in the mass media and the social debate. The contending parties also used markedly asymmetric discourses – one based on political principles of equality and non-discrimination, the other on biological and religious arguments.

Equality and difference

Made in the name of equality, the LGBT movement's demands were all or nothing, as expressed in the marriage campaign motto, "*the same rights with the same names*". Its goal was not the recognition of difference but of equality and dignity (or, better, of equal dignity), hence its efforts to impose the labels of "equal marriage" and "marriage for all" over that of "gay marriage". Dignity, explained Pedro Zerolo, "*is having the same rights, the same duties and with the same names as heterosexuals. Because if they don't have the same names, that's apartheid*" (cited in *P/12*, 16 July 2006).

Even the specific demand of difference recognition that eventually crystallised into the Gender Identity Law was based on the argument that such recognition is what equality required. As trans activist Diana Lavalle emphasised, "*we are not asking for anything that every other person does not have, we do not want anything taken from anybody ... The only thing that we want is equality*" (cited in *El Puente* 56, 1 June 2012). Action was needed to guarantee trans persons the enjoyment of a basic right – the right to identity – that their fellow citizens took for granted, and without which access to all other rights remained closed. "*With an ID that does not reflect their real identity, a person is not accepted in school, by the health system, nor has she the possibility of voting*" (Esteban Paulón in *P/12*, 29 September 2010). "*If the Gender Identity Law is passed, I am going to be able to go to a hospital with no fear, to request a loan or use my credit card; [because] when I show it in a store they often think that it is stolen*", explained a trans college student (*P/12*, 29 July 2010).

Human rights and democracy's debts

In Argentina, the topic of identity struck a particularly sensitive chord. Well into the 2000s, the plight of the Grandmothers of Plaza de Mayo remained very much alive as they kept looking for their grandchildren, who had been kidnapped along with their parents under the dictatorship, illegally adopted and deprived of their identities. Berkins stated it clearly when anticipating that "*for the first time our community can achieve recognition, in a country where the wounds caused by theft and suppression of identity are still open*" (*P/12*, 17 April 2012). Unequal access to human rights was therefore defined by trans activists as one of the main debts of democracy:

> "Democracy is in debt with us ... We have historically been excluded and we have stayed under the poverty line. We are undoubtedly democracy's forgotten ones." (Marcela Romero in *LN*, 17 April 2012)

Great emphasis was placed on the symbolic dimension of recognition. This demand was raised not just by the trans community but by the LGBT movement as a whole. The most basic of the claimed rights was indeed the right to exist, to be recognised as a subject of rights and counted as a member of the community. "*A person who has no identity does not exist, and in Argentina we did not exist*", explains Marcela Romero (interview at Puntal.com.ar 11 may 2012). Within the discursive frame of human rights, the gender identity law was apprehended as "*a door of access to all other rights*" (Diana Lavelle, interview in *El Puente* 56, 1 June 2012). Therefore, the campaign focused on demonstrating the consequences that the lack of trustworthy identity papers had on the chances of exercising and enjoying citizenship rights – including political rights, the basis of democratic legitimacy.

In fact, all of the LGBT movement's demands were framed in the language of citizenship and human rights. Regarding the right to marry, for instance: it was systematically emphasised that much more than a ceremony and a party were at stake, as a huge amount of rights were associated to it – health insurance coverage, inheritance, pension after a partner's death, legal residence and the right to citizenship when one of the partners was foreign-born, joint acquisition of equity, legal protection and support in case of a divorce, joint adoption and transference of rights to children.

Having embraced the human rights discourse, it was only natural for the LGBT movement to draw analogies between their own struggle and those of other emancipation movements of the past, such as women's or African-Americans', as well as between the proposal of a separate legal institute for homosexuals and the South African apartheid regime. In the same vein, identification with persecuted Jews was outlined in the context of the congressional debates in which LGBT activists were the targets of offensive expressions that Argentine law would not have allowed if directed against Jews.

Given the political efficacy of the human rights discourse, the opposition to equal marriage also tried to appropriate it to legitimise its positions. ACIERA, a network of evangelic churches, proclaimed the existence of a contradiction between the proposed reform and the principles contained in the Universal Declaration of Human Rights, the International Covenant on Civil and Political Rights and the Inter-American Convention on Human Rights. Legislators and activists mobilised against same-sex marriage aimed to appear as standard-bearers of the human rights of children – whose healthy development, they argued, depended upon their being part of a so-called normal family with a mum and a dad – and to spark a controversy around adoption. This attempt was repelled by LGBT activism with two arguments: that the new law would not modify the current adoption regime, which already enabled adoption by homosexual individuals; and that homoparental families already existed and all that was asked is that they be granted equal legal recognition and protection. A lesbian activist and mother of triplets put it as follows:

> "We are not asking for permission to start a family: our family already exists. What we are asking for is that our children have the same rights as the children of any other family." (Silvina Maddaleno, interview in *P/12*, 15 November 2009)

Real people vs. abstract concepts

Throughout the debate, opponents to equal marriage insisted that what was truly at stake was the concept of marriage and that reality needed to fit into the definition and etymology of the word. The LGBT organisations, in turn, tried to show that it was the rights, well-being, and happiness of real people what was at stake. Therefore, their arguments in terms of rights were typically accompanied by first-person descriptions of the brutal impact caused by the lack of rights.

The media contributed with the plentiful diffusion of life stories with an emphasis on injustice, suffering, perseverance, and love. The protagonists of those stories looked like they had been born to be on television: middle class youth, pleasant and educated; professional women, the mothers of beautiful children; nice old ladies whom anyone would like as neighbours.

A similarly powerful effect was produced by two key interventions that took place before the congressional votes. The first was a speech by a socialist representative who identified himself as the father of a gay son and protested against the injustice that his homosexual son did not have the same rights as his heterosexual one. According to Augsburger, this speech "*helped change several wills*" (*P/12*, 6 May 2010). In turn, just before the Senate vote a letter was made public that a teenager had addressed to the senators, in which he told them about his (perfectly normal and happy) life as the son of a gay father, of whom he declared to be proud and to whom he offered his wholehearted support (*P/12*, 14 July 2010).

Conclusion: a new egalitarian common sense?

The speed of change surprised both participants and observers. The struggle for equal marriage and gender identity, however, began well before the bills were introduced. The processes to get these laws passed, laborious in themselves, were the result of a much longer process of organisation and visibilisation of the community, discourse production, and articulation of alliances. As a result, what had been unthinkable just a decade earlier eventually became real. In the early 2000s, the maximum realistic aspiration was that of extending civil union to the national level; marriage was outside the realm of the possible. Even in the mid-2000s the idea that the public health system could provide sexual reassignment surgeries –at the time still banned – seemed like fantasy. So fast were the changes that today:

> "it is difficult for the very young people to get to understand the suffering of gays, lesbians and transvestites just ten years ago. Luckily." (Marcelo Suntheim, interview in *El Puente* 17, 2 September 2011)

What made this change possible? Key to the movement's success were its unity –which translated into a considerable ability to overcome internal differences – and its diversity – the fact that it comprised a variety of organisations representative of specific subgroups, focused on different topics and capable of turning into the visible face of each of the movement's demands. It is worth noting the trajectory – from victimisation to action – followed by the trans community, which now has its own organisations led by transsexuals and transvestites, its own agenda, and a demonstrated ability to work on behalf of their target population.

An additional factor was the movement's considerable skill in building alliances with other movements (the women's movement firstly, but also workers' and students') and to gather the support of a wide spectrum of actors holding various forms of social capital: academics, researchers, scientists, artists, journalists. Also important was the deep transformation undergone over the previous decades by the self-definition of the political left and, consequently, by its view of sexual diversity. In marked contrast with the leftist ethos of three decades ago, it is now common sense indeed that "*you cannot be on the left and be sexist or xenophobic ... racist, homophobic*" (Pedro Zerolo, interview in *P/12*, 16 July 2006).

The formation of a social consensus around the LGBT demands was impelled by at least two main factors. One was its framing of the claim in the language of human rights, within a national context that was highly sensitive to such appeals as well as relatively secularised. A convincing articulation of this discourse even allowed the LGBT movement to appropriate a topic most dear to conservatives: the defence of family and child. The second factor is the role played by the mass media, and particularly by television. Journalistic coverage was mainly favourable to the movement and awarded an unprecedented visibility to the community, via stories of love and suffering and memorable debates – in which another asset of the movement also appeared: well-prepared, articulate leaders and activists. This was how debate and conversation on the issues involved entered the home.

Furthermore, the conversion of social consensus into legal mandate was possible thanks to the presence of a set of (mostly female) allies who played key roles in the legislative, executive and judicial branches of government. Finally, a virtuous circle seems to have been formed as a result of interactions between cultural and legal change. Indeed, for legal change to start in the first place, "*there had to be a [prior] change and an opening; even many people from LGBT organisations were surprised at the social consensus that existed – at least in the big metro areas*" (Santiago, interview 6 August 2012). The exhibition of social consensus was key in getting the political system moving towards legal change.

In turn, the increased visibility generated by the new laws seems to have further changed the social perception of homosexuals and transsexuals; in that sense, the effects of legal change spread well beyond those who made or intend to make direct use of the new institutions and benefits:

> "The [equal marriage] law opened ... a window of opportunity for a change in social customs and traditions. It brought the opportunity for some people to talk at home. And if they can get married, why wouldn't they be able to walk on the streets holding hands? This law continues to create opportunities ... for more acceptance, more recognition ... [Many people] came out of the closet during the past one and a half years." (Santiago, interview 6 August 2012)

As a result of the educational dimension of the law, "*the new generations will grow up within the normalcy of equal marriage, just as today they do so regarding female vote*", predicts the CHA president (*CL*, 9 August 2010). Meanwhile, several tasks remain: monitoring the implementation of the new laws; promoting public policies allowing for the translation of legal equality into equality of treatment and opportunity; claiming other rights included under the umbrella of sexual rights (especially that of abortion, demanded by the LGBT movement's main allies); and fighting against other forms of inequality and discrimination. Besides the usual obstacles, though, an additional one conspires against getting the job done: the movement's demobilisation and loss of unity that was the result of its great victory.

> "Demobilisation comes naturally after you have achieved your highest goal ... [When you further a new demand] politics itself tells you 'well, your community already got almost everything ...' ... Society itself tells you 'OK, it's done, stop.' ... The other day a cab driver asked me: 'What else do you want now?'" (Ignacio, interview 1 August 2012)

Note

1. Except for those of public figures, the names of the interviewees were changed.
   Material for this paper came from a variety of sources, such as:
   - *Clarín* (*CL*, http://www.clarin.com/), *Infobae* (*IB*, http://www.infobae.com), *La Nación* (*LN*, http://www.lanacion.com.ar/), *Página/12* (*P/12*, http://www.pagina12.com.ar), *Perfil* (*PL*, http://www.perfil.com), *Tiempo Argentino* (*TA*, http://tiempo.infonews.com).
   - Radio broadcast *El Puente*, Facultad de Ciencias Sociales, Universidad de Buenos Aires. Available at http://programadecapacitacion.sociales.uba.ar/elpuente/
   - Websites: 100 per cent Diversidad y Derechos (http://100porciento.wordpress.com); Asociación de Mujeres Meretrices de Argentina (AMMAR; http://www.ammar.org.ar); Asociación de Travestis, Transexuales y Transgéneros de Argentina (ATTTA, http://www.attta.org.ar); Comunidad Homosexual Argentina (CHA, http://cha.org.ar); Cristianos Gays (http://www.cristianosgays.com); Federación Argentina LGBT (FALGBT, http://www.lgbt.org.ar), Frente Nacional por la Ley de Identidad de Género (http://frentenacionaleydeidentidad.blogspot.com.ar/); La Fulana (http://www.lafulana.org.ar); Fundación Buenos Aires SIDA (http://www.fbas.org.ar); Lesmadres (http://www.lesmadres.org.ar); Marcha del Orgullo (http://www.marchadelorgullo.org.ar); Nexo (http://www.nexo.org); Vox (http://www.voxargentina.org).
     Interviews: Esteban Paulón (July 31, 2012); Alejandro Freyre (August 4, 2012); José María Di Bello (August 4, 2012); *Ignacio* (August 1, 2012); *Mariano* (August 2, 2012); *Santiago* (August 6, 2012); *Paula* (August 14, 2012).

References

Aldao, M., and L. Clérico. 2010. *Matrimonio Igualitario. Perspectivas sociales, políticas y jurídicas*. Buenos Aires: Eudeba.

Álvarez, A. G. 2009. "De figura performativa a identidad narrativa. Transformaciones en la construcción de las subjetividades travestis en tiempos neoliberales." Paper presented at the 28th LASA International Congress. Rio de Janeiro, June 11–14, Brazil.

Bimbi, B. 2010. *Matrimonio Igualitario. Intrigas, tensiones y secretos en el camino hacia la ley*. Buenos Aires: Planeta.

Congregación para la Doctrina de la Fe. 2003. "Consideraciones acerca de los proyectos de reconocimiento legal de las uniones entre personas homosexuales." Accessed April 26, 2013. http://www.zenit.org/article-9793?l=spanish.

Fígari, C. 2012. "Gay Marriage/Matrimonio Igualitario in Argentina: From Conservative Claim to Progressive Demand." Paper presented at the 30th LASA International Congress, May 23–26. San Francisco, CA.

Jones, D. 2010. "Discursos evangélicos sobre cuestiones LGBT en la Argentina contemporánea (2000–2010)." Paper presented at the 29th LASA International Congress, October 6–9, Toronto, Canada.

Salinas Hernández, H. M. 2012. "Matrimonio Universal en la Ciudad de México: Análisis de la política pública." Paper presented at the 30th LASA International Congress, May 23–26, San Francisco, CA.

The author

Inés M. Pousadela is an Argentinean political scientist now residing in Washington, DC. Former professor at the University of Buenos Aires and visiting researcher at Georgetown University and the University of Maryland, she is currently a research fellow at American University and an independent political consultant for NGOs and international organisations. She has published books on political representation and participation in Argentina and Latin America, as well as a number of articles and book chapters on political theory, democratisation, social mobilisation, representation, political culture, and the issues of corruption, transparency, and accountability.

The emergence and re-emergence of civil society: a brief history of civil society in Europe, from the Magna Carta to the Eurozone crisis

Brian Pratt and Rowan Popplewell

Since 2008, Europe has been mired in a series of complex and interconnected economic, financial, fiscal, and political crises, prompting citizens across Europe to erupt onto the streets to protest against cuts to pensions and public services, taxation increases, and government corruption. The current crises have resulted in profound challenges for many civil society organisations, yet they also arguably present an opportunity for many groups and have prompted the re-emergence of a more vibrant and politically engaged civil society in some European countries, as compared to the large contract delivery NGOs which have dominated recent years, sometimes to the detriment of small civil society organisations which are the backbone of European civil society However, in order to fully understand these current and emerging trends, we should cast our gaze further back. This article aims to contextualise recent trends through providing a brief history of the emergence and development of civil society in Europe, drawing on seven longer case studies from the UK, Republic of Ireland, the Netherlands, and Greece. It will then discuss some of the main issues and trends, and reflect on possible future directions for civil society in Europe.

L'émergence et la ré-émergence de la société civile : un bref historique de la société civile en Europe, de la Magna Carta à la crise de l'Eurozone
Depuis 2008, l'Europe est enlisée dans une série de crises économiques, financières, fiscales et politiques complexes et reliées entre elles qui poussent les citoyens des quatre coins de l'Europe à descendre dans la rue pour protester contre les coupes sombres opérées dans les retraites et les services publics, les hausses des impôts et la corruption des gouvernements. Les crises actuelles ont abouti à de profondes difficultés pour de nombreuses organisations de la société civile, mais on peut aussi soutenir qu'elles présentent une occasion pour de nombreux groupes et qu'elles ont suscité la ré-émergence d'une société civile plus dynamique et politiquement engagée dans certains pays européens, par rapport aux grandes ONG chargées de mettre en œuvre des contrats, qui ont dominé au cours des dernières années, parfois aux dépens des petites organisations de la société civile qui constituent la clé de voûte de la société civile européenne. Cependant, afin de comprendre pleinement ces tendances actuelles et émergentes, nous ferions bien de revenir un peu en arrière. Ce Point de vue cherche à contextualiser les tendances récentes en fournissant un bref historique de l'émergence et du développement de la société civile en Europe, en s'inspirant de sept études de cas effectuées sur des périodes assez longues au Royaume-Uni, en République d'Irlande, aux Pays-Bas

et en Grèce. Il traitera ensuite de certaines des principales questions et tendances, et réfléchira à des directions futures possibles pour la société civile en Europe.

El surgimiento y re-surgimiento de la sociedad civil: una breve historia de la sociedad civil en Europa, de la Carta Magna a la crisis de la eurozona

Desde 2008, Europa ha estado sumergida en una serie de complejas e interconectadas crisis económicas, financieras, fiscales y políticas. Las mismas han llevado a que los ciudadanos de toda Europa irrumpan en las calles y protesten contra los recortes a las pensiones y a los servicios públicos, contra los aumentos a los impuestos y contra la corrupción gubernamental. Las crisis actuales han implicado retos para diversas organizaciones de la sociedad civil. Sin embargo, puede decirse que para muchos grupos también representan una oportunidad, toda vez que han impulsado el re-surgimiento de una sociedad civil más vibrante y politizada en varios países europeos, en comparación con las grandes ONG orientadas a la entrega [de bienes y servicios] por contrato. En años recientes, dichas ONG han dominado el espacio público, perjudicando en ciertos momentos a las pequeñas organizaciones de la sociedad civil que constituyen la espina dorsal de la sociedad civil europea. Para comprender mejor estas tendencias actuales y emergentes, es necesario revisar la historia. Apoyado en siete largos estudios de caso realizados en el Reino Unido, la República de Irlanda, Holanda y Grecia, este artículo pretende contextualizar las tendencias recientes a través de una breve historia del surgimiento y del desarrollo de la sociedad civil europea. Asimismo, analiza algunos de los principales temas y tendencias, presentando ciertas reflexiones en torno a los posibles caminos que se abren para la sociedad civil europea de cara al futuro.

O surgimento e ressurgimento da sociedade civil: uma breve história da sociedade civil na Europa, desde a Carta Magna até a crise da Zona do Euro

Desde 2008, a Europa tem enfrentado uma série de crises econômicas, financeiras, fiscais e políticas que são complexas e interconectadas, estimulando cidadãos de toda a Europa a irem para as ruas protestar contra cortes na previdência e nos serviços públicos, aumento nos impostos e corrupção no governo. As crises atuais resultaram em desafios profundos para muitas organizações da sociedade civil, embora elas também possam representar uma oportunidade para muitos grupos e têm estimulado o ressurgimento de uma sociedade civil mais vibrante e politicamente engajada em alguns países europeus, comparado a ONGs que implementam grandes contratos, que têm dominado os últimos anos, às vezes em detrimento de pequenas organizações da sociedade civil, que são a base da sociedade civil europeia. Porém, para compreender plenamente estas tendências atuais e emergentes, devemos voltar nossos olhos atentamente para o passado. Este Ponto de Vista tem como objetivo contextualizar tendências recentes, oferecendo um breve histórico do surgimento e desenvolvimento da sociedade civil na Europa, baseando-se em sete estudos de caso mais longos do Reino Unido, República da Irlanda, Países Baixos e Grécia. Ele então discutirá algumas das questões e tendências principais e refletirá sobre possíveis direções futuras para a sociedade civil na Europa.

Civil society in Europe: a very brief history

Taking an historical perspective when analysing contemporary civil society provides insight into how the key roles and functions of civil society developed over time, and how many of the current issues have roots that extend back centuries. This section draws particularly on case studies from the UK (Savage and Pratt 2013; Soteri-Proctor, Phillimore, and McCabe 2012; Howard and Pratt-Boyden, this issue).[1]

The emergence of civil society: The Middle Ages to the First World War

Many European countries have rich and centuries-long civil society traditions, forged and shaped by their relationship to the state. In the UK, the genesis of civil society can be traced back as far as the Middle Ages and the signing of Magna Carta in 1215: this was the first formal document enshrining civil liberties in law and ensuring the balance of power between the King and his subjects. Civil society and citizens were considered to be the principal source of power, which was then offered to the state on a contract, allowing the state to rule in return for providing services such as physical security and the protection of property rights. Many of the political functions of civil society that emerged at this time – generating the social basis for government, promoting political accountability, and supporting the rights of citizens and the concept of citizenship – have survived until the present day. Indeed, civil society in the UK also has a long tradition of holding the Government and other powerful groups in society to account through advocacy, lobbying, and campaigning. Groups lobbied those in power for special interests, but also in support of more altruistic and progressive causes, the most famous being the anti-slavery lobby in the late eighteenth century. Indeed, the campaign for the abolition of slavery was one of the world's first transnational civil society movements. Supporters on both sides of the Atlantic put pressure on the British Navy to stop the trade where it could, lobbied parliaments, and successfully instituted a boycott of sugar produced by slave labour. This was followed in the nineteenth and early twentieth centuries by movements against child labour and for women's suffrage, amongst many other important issues.

There is also a long history of service provision among civil society in the UK. The first social and welfare services in England were either provided by the Church or through patronage based on feudal relationships. In the sixteenth century the state began to see itself as having some responsibility for poverty alleviation, at least towards what it regarded as the deserving poor. At this time, the state's primary role was encouraging and regulating private and locally-based provision through parishes, churches, and other society-based institutions rather than directly providing services itself.[2] The first charity laws were passed by Elizabeth I in 1601, and listed a set of purposes and activities that were deemed by the state to be of general benefit to society and thereby deserving of funding. Thus, most services would have been based in local communities and provided by a mix of religious institutions and benevolent feudal landlords – something probably true across most of Europe where some of the oldest surviving institutions are welfare and educational establishments.[3] By the nineteenth century there was a major input into educational, health, and other social services by civil society groups as well as the development of new forms of civil society including trade unions, mutual societies, and cooperatives. This growth in service provision was encouraged by the British state, which continued to believe it was responsible for regulation rather than the provision of these activities, leading to the emergence of new laws and regulatory mechanisms.

The growth of the welfare state: the First World War to the 1980s

The early twentieth century in Europe saw the beginning of a growth in state involvement in many areas of social services. Although some of these changes began before the First World

War, the destruction caused by war and its aftermath, particularly the economic crisis that engulfed Europe and North America, led to an increased concern in both the positive economic benefits of improved services but also the fear of uprisings from a disaffected and unemployed populace. This formed the basis of Europe's welfare states – a form of government that intervenes to protect and promote the economic and social welfare of citizens through legislation and service provision – which were rolled out across Europe following the Second World War.

In some areas of Europe, the expansion of state-led services weakened the role of civil society organisations as providers of social services. In the Netherlands prior to the introduction of the welfare state, social service provision was structured around a system known as pillarisation, where social services were segmented along vertical cleavages, principally by religious denomination. This meant that each religious community had its own set of services, typically funded by the state but provided by members of the community for others within that religious denomination, creating duplicate services for the different sections of society that grew in both size and number as government funding increased. The introduction of the welfare state, and the absorption and consolidation of services by the state that this entailed, gradually removed these duplications in service provision, and in doing so fundamentally weakened Dutch civil society and the religious communities that had previously provided these services.

Elsewhere in Europe, religious institutions maintained their control over many services but relied increasingly on subsidies from the state. In the Republic of Ireland, the long running fight for independence from the United Kingdom as well as the strength of the Catholic Church meant that the Church historically provided many social services as opposed to the government of the day. After independence, the role of the Catholic Church and other civil society groups in the provision of services continued to grow as state funding for services increased.

Rolling back the state: the 1980s to the 2000s

Since the 1980s, in part due to recession but also to an ideological reaction against what has been seen as an overbearing dominance by the state, the original idea of the state caring for people "from the cradle to the grave" has been contested and the welfare state rolled back across Europe. Initially the reaction was to privatise services previously carried out by the state directly. The return to private service-provision prompted a series of experiments across Europe with privatising schools and health and institutional care facilities, although usually with some degree of state oversight.

The reduction in state-led service provision initially led to some increase in the income of civil society groups dedicated to the provision of services on contract, taking their incomes well beyond previous levels gained from individual donations and private philanthropy. However, the massive expansion of the service delivery function of many of these organisations led to concerns regarding dependency on state funds, resulting in declining autonomy and ability to hold the state to account. This has led to the perceived co-optation and pacification of sections of civil society in some European countries. For example, in the Republic of Ireland high levels of state funding for service delivery functions of civil society, the introduction of formalised funding regimes, and the institution of a form of corporatism known as social partnership that governed state-society relations, allowed the state to co-opt large sections of civil society, resulting in a somewhat politically neutralised civil society (Popplewell 2013). A similar story exists in the Netherlands, where civil society has historically been subsidised by the state in order to pacify tensions in society. Concerns remain that government financing of civil society organisations has increased state control over these organisations and restricted their independence (Habraken et al., this issue).

The Eurozone crisis and the re-emergence of civil society: challenges and opportunities

In 2008 several European countries entered recession. Precipitated by the bursting of the US housing bubble in 2007, the effects quickly spread across the Atlantic, sparking a series of interconnected economic, financial, fiscal, and political crises across Europe that continue to cause uncertainty, particularly over the future of the Euro. While this has created complex, multiple, and profound challenges for citizens and civil society organisations, it has also arguably created opportunities for civil society and prompted the rediscovery of widespread civic action and popular protest. However, not all civic action that has emerged in response to the current crisis can be described as progressive; in many European countries populist, violent, fascist, racist, xenophobic, and ultra-nationalist groups and movements have grown alongside their more progressive counterparts.

Challenges to formal and established civil society organisations

The multiple crises that have hit Europe in recent years have brought many formal and established civil society organisations to a crossroads. Across Europe, many organisations have experienced a profound and sustained funding crisis. While this primarily results from cuts to state funding, declining consumer income and rising unemployment have also put pressure on funds generated from individual and corporate donors. After decades of operating in an environment characterised by economic growth and ever increasing resources, civil society organisations are now facing tough choices about how to best fund their activities in a context of declining resources. Many civil society organisations are attempting to ride out the storm through economising – primarily through cutting staff costs, activities, programmes, and services – and through diversifying their funding base. However, cutting expenditure and searching for alternative pots of money to bridge the gaps created by cuts to state funding may not be enough to keep many civil society organisations open and active. As the overall pool of resources has shrunk, competition has increased, resulting in an ever more squeezed and competitive market place. Many organisations, particularly smaller ones and those historically reliant on state funds with underdeveloped fundraising arms, are simply not able to compete against bigger organisations with greater fundraising capacity.

In addition to this funding crisis, many established civil society organisations are also experiencing a crisis of relevance. In many countries, they have faced increasing criticism from other elements of civil society about their ability and legitimacy to act on behalf of citizens and respond to the current crisis. In Greece, these organisations have been criticised by many activists for being unsupportive of citizen-led movements seeking social and political change, and of being undemocratic and unaccountable, irrelevant and even complicit in creating and supporting a system "*that provides symptomatic relief to social problems so that it doesn't have to challenge real power imbalances*" (Pantazidou, this issue).

These dual crises raise questions about the ability of established civil society organisations to respond in progressive and innovative ways. However, while many organisations are facing severe challenges, it could also be argued that opportunities are opening up for civil societies that have historically been co-opted and pacified by governments across Europe. The reduction in state funding in particular may present an opportunity for a sector that has become increasingly depoliticised; as organisations become less reliant on state funds they may become increasingly financially and politically independent, particularly if they can find support from elsewhere in civil society. For example, a number of non-profit organisations in the Netherlands have responded to cuts in state funding by re-aligning themselves within civil society, and now draw support from within civil society as opposed to the state, increasing their independence and ability to hold the government to account (Habraken et al., this issue). Conversely, in the Republic of Ireland, where many organisations remain heavily dependent on the Irish state despite substantial cuts to state funding, significant questions remain about the political independence of civil society and its ability to respond to the current crises in a creative, innovative, and progressive manner (Popplewell 2013).

The re-emergence of social movements and civil society activism in Europe

Across Europe, new social movements and civil society organisations have sprung up in response to the financial crisis, from the Syntagma Square movement in Greece where thousands of civil society activists occupied one of the main squares in Athens for more than two months, to the Occupy groups that have been active in many European cities. Form is important to both these movements: they aim to create a new political culture through their chosen modality of citizen organisation. Occupy sees itself as an example of "*pre-figurative political action*" that seeks to create a miniature replica of the wider society it wishes to see emerge (Howard and Pratt-Boyden, this issue), whereas the Syntagma Square movement is based on a model of direct democracy that puts "*horizontality and autonomy*" at the core of both their mode of organising and their vision for a different society (Pantazidou, this issue).

A challenge for both these movements has been how to ensure the sustainability of citizen action once the initial protest tails off. In Greece, a "*polymorphous assemblage of citizen initiatives*", from community run soup kitchens to neighbourhood assemblies, has replaced the mass participation events such as the Syntagma Square occupation of 2011 and the Greek riots of 2008 that initially got Greek citizens onto the streets (Pantazidou, this issue). However, this has not been replicated elsewhere; in many other European countries very few social movements have managed to harness initial outbursts of citizen energy and channel it in productive and sustainable ways towards the attainment of wider political and economic alternatives. Although these movements are often inherently critical of the more established civil society organisations, it is possible they can learn something from them in this regard, just as established civil society organisations have a lot to learn from these newer, and often more progressive and innovative, forms of citizen action.

"Under the radar" civil society organisations

Another set of organisations that are prevalent, though scarcely recognised, in European civil society are small, informal, grassroots, or "under the radar" organisations. These organisations have a long history throughout Europe and make up the largest component of civil society (Soteri-Proctor et al. 2012). It is at this level where civil society is the most organic in formation, functioning, and in adapting to the changes in society. They have gained prominence in recent years partly as a result of a growing interest among policymakers in the role of citizens and local communities in identifying and resolving social and political problems, and partly as a way of filling the gaps in service provision created by cuts to public services.

Consequently, many of these informal organisations are being put under pressure to scale up their activities and become more professional and formal. The Makassar Neighbourhood Community in the Netherlands is facing this exact issue, but some in the community are concerned that professionalisation will mean losing the vibrancy, creativity, and inspiration that stem from the organisation's informality, and they risk falling into "*the trap of becoming part of the formalised pacifying, docile bureaucratic system, limited by too many rules and regulations and thus risking its unique bridging position and rootedness in the neighbourhood*" (Habraken et al., this issue).

Uncivil society in Europe

However not all civic action and civil society groups that have emerged in response to the financial and economic crises can be described as progressive. Many civil society demonstrations against austerity measures have turned violent, including student anti-tuition fee protests in London in November 2010 and the Greek riots of 2008. Further, ultra-left and far-right political and civil society organisations have also gained ground in recent years. Throughout Europe, fascist, racist, xenophobic, and ultra-nationalist sentiments and organisations are on the rise and are feeding a very different kind of citizen action. For example, in Greece far-right citizen groups have also set up community run soup kitchens and other charitable initiatives that cater only to Greek citizens and which, because of their military-like hierarchy, operate more quickly and efficiently than their democratically organised counterparts (Pantazidou, this issue).

What next for civil society in Europe?

At the time of writing, uncertainty still hangs over the future of the Eurozone. Whether the crisis continues to spread through Europe and what the implications of a collapse of the Euro might be are difficult to gauge. What is certain is that the dominant role of the state is being questioned in many countries and by default the roles of civil society are also being reviewed. Whereas the first phase of this change was to contract out services to civil society assisted by state funding, there is now an emphasis on communities absorbing the costs of many local services, often resulting in massive financial cuts to civil society organisations throughout Europe. By adopting a historical perspective we can see that the current retreat of the state and re-emphasis on the service delivery functions of civil society is actually part of a longer term trend that has been exacerbated and entrenched by the current crises.

What could be described as new is the widespread re-emergence of street demonstrations across Europe and the re-politicisation of elements of civil society. Some civil society organisations feel that they were led astray, compromising their independence too much through their financial dependence on the state, and are moving towards a more political, advocacy-based approach. The refocusing on the more political functions of civil society by many civil society organisations, groups, and movements throughout Europe has arguably prompted the re-emergence of a more vibrant, dynamic, and healthy civil society. However, a number of questions remain regarding the ability of civil society, especially many established civil society organisations, to respond in progressive and innovative ways to the current crisis and the sustainability of civic action; two significant crossroads for civil society in Europe.

Related to this is the ability of more progressive elements of civil society to mount an effective challenge to the simultaneous re-emergence of uncivil society groups throughout Europe. Indeed there is a lot going on across Europe that has not been captured either in this paper or the case studies upon which it is based. We believe that the limited examples discussed here point to a much broader picture regarding the rise of fascist, racist, xenophobic, and ultra-nationalist sentiments and organisations throughout

Europe, which have the very real potential to push political and civil society in dangerous directions. Consequently, halting and reversing the rise of uncivil society must be a key concern for civil society actors across Europe when deciding how to move forward from the current crisis, and represents possibly the most important crossroad for mainstream European civil society today.

Notes

1. Case studies from the UK include Savage and Pratt 2013; Soteri-Proctor, Phillimore and McCabe 2012; Howard and Pratt-Boyden, this issue. Case studies from the Netherlands include Habraken et al., this issue; Makassar case study. Other case studies include Popplewell 2013; and Pantazidou, this issue. Primarily we are referring here to Western Europe, as the experience of Central and Eastern Europe, particularly following the Russian Revolution and the formation of the old Warsaw Pact is very different.
2. There is evidence to show that the state did fund some charities in the UK even at this time, predating the massive increase of state funding for civil society organisations by several hundred years (as in the UK case study).
3. The King's School Canterbury, founded in 597, and re-founded in 1547 claims to be one of the oldest still operating in the UK. In Italy there are hospitals dating back over a thousand years, some of which are not just for medical treatment but general charitable foundations running several services including orphanages, and general welfare funds.

References

Popplewell, R. 2013. *Briefing Paper 35: Responding to crisis: Understanding the effects of political and economic crisis on civil society in the Republic of Ireland.* Wallingford, UK: INTRAC.

Savage, O. and B. Pratt. 2013. *Briefing Paper 38: The history of UK civil society.* Wallingford, UK: INTRAC.

Soteri-Proctor, A., J. Phillimore and A. McCabe. 2012. *Under the radar: Civil society at the grassroots in the UK.* New Delhi: PRIA.

The authors

Brian Pratt is one of the founding members of INTRAC and is the Executive Director, as well as the Editor-in-Chief of *Development in Practice.* He has worked in development since 1973, initially at the University of Cambridge and in Peru on the political economy of Peru. Brian completed his PhD at Cambridge University in 1980. He was also a Field Director for Oxfam in the Andean Countries (1977–82), before returning to the UK to develop Oxfam programmes and Departments of Research, Evaluation, Development Policy, and Information Systems (1982–90). During this period he wrote the *Oxfam Field Director's Handbook*, and was the founding editor of *Development in Practice* and the Development Guidelines series of seven books. The primary focus of Brian's publications and consultancies is on strategic policy issues for NGOs and Monitoring and Evaluation (M&E).

Rowan Popplewell (corresponding author) works as a Research Officer for INTRAC's Research Department, primarily on INTRACs NGO Research Programme. Her current research interests include action research, research in NGOs, trends in monitoring and evaluation, and the impact of political and economic crises on CSOs. Rowan has also worked at the Scottish Catholic International Aid Fund (SCIAF) as a Policy Officer, and completed an MSc in Violence, Conflict, and Development from the School of Oriental and African Studies (SOAS), University of London, where she focussed on NGOs and civil society in conflict-affected and post-conflict areas. Her areas of geographical expertise include Central and Eastern Africa and South Asia.

Occupy London as pre-figurative political action

Neil Howard and Keira Pratt-Boyden

This paper reflects on the Occupy London encampment as an example of pre-figurative political action. The paper articulates the major media-political criticism of Occupy as "unwieldy" and "in search of a narrative" (Anonymous 2012), before drawing on case studies of two of the Occupation's components – the Tent City University and the Welfare and Well-being Group – to suggest that this criticism fails to understand the intentional choice that "directionlessness" represented. Indeed, the refusal to adopt formal leadership or any pre-defined ideological goal was a conscious strategy which pointed towards the kinds of possible future(s) desired.

Occupy Londres en tant qu'action politique préfigurative
Cet article traite du camp d'Occupy Londres comme exemple d'action politique préfigurative. L'article formule l'importante critique médiapolitique d'Occupy comme étant « complexe » et « en quête d'une dimension narrative » (Anonyme 2012), avant de se servir d'études de cas de deux des éléments de l'Occupation – la Tent City University et le Welfare and Well-being Group – pour suggérer que ces critiques ne saisissent pas le choix intentionnel que représentait ce « manque de direction ». De fait, le refus d'adopter un leadership formel ou tout but idéologique prédéfini constituait une stratégie consciente qui indiquait les sortes d'avenir(s) possible(s) souhaités.

Occupy Londres como una acción política prefigurativa
El presente artículo analiza el campamento de Occupy Londres como un ejemplo de acción política prefigurativa. Se examinan las principales críticas realizadas por los medios en el sentido de que Occupy era "poco manejable" y andaba "en busca de una narrativa" (Anonymous 2012). Asimismo, se revisan dos estudios de caso realizados con sendos componentes de la Ocupación: la Universidad de la Ciudad de las Carpas y el Grupo de Bienestar, los cuales evidencian que las críticas efectuadas no comprendieron lo que representaba haber decidido intencionalmente no tener dirección política. De hecho, el rechazo a adoptar liderazgos formales o a tener objetivos ideológicos predefinidos constituyó una estrategia consciente que apuntó hacia ejemplos de posible(s) futuro(s) deseado(s).

Occupy London como ação política pré-figurativa
Este artigo reflete sobre o acampamento "Occupy London" como exemplo de uma ação política pré-figurativa. O texto articula a grande crítica política da mídia à Occupy como sendo "descoordenado" e "em busca de uma narrativa" (Anônimo 2012), antes de basear-se em estudos de caso de dois dos componentes da Ocupação – a Tent City University e o Grupo sobre Bem

Estar – para sugerir que esta crítica não consegue compreender a escolha intencional que "a falta de direção" representou. De fato, a recusa em adotar uma liderança formal ou qualquer objetivo ideológico pré-definido foi uma estratégia consciente que apontou em direção aos tipos de futuro(s) possível(is) desejado(s).

Introduction

On 15 October 2011, hundreds of people gathered nervously on the steps of St Paul's Cathedral, responding to a "call to action" paralleled in cities across the globe. By the morning of 16 October, Occupy London and its camp were born, and for the next four months would form the central focus of national media attention and the nucleus of national political protest. In this paper, we examine the emergence and experience of this camp and suggest that, in contrast to commonplace critiques besmirching it as directionless or frivolous, it in fact represented an example of consciously-conceived, pre-figurative political action. We will make our case through an analysis of the two activities with which we were most deeply involved – Tent City University, and the Welfare and Well-being Working Group.

The emergence of Occupy London

It is difficult to accurately trace or date the genesis of any social event. All causal chains are explanatory, narrative devices that help make sense of the world but which, in so doing, inevitably simplify it (Tilly 1997, 2002). This caveat notwithstanding, it is possible to identify a broadly linear proximate backdrop to the emergence of the Occupy London protest camp which took up residence outside the London Stock Exchange in October 2011.

This backdrop begins with the onset of the global financial crisis marked so spectacularly by the collapse of the Lehman Brothers investment bank in late 2008. When the US housing bubble edged towards bursting under the weight of sub-prime mortgages and the corresponding rise in foreclosure rates in 2006/7, the stability of the previously unquestioned global financial pyramid began to look increasingly ephemeral. Mortgage-backed securities turned toxic, investors and bond-holders realised the worthlessness of so many of their assets, and in response credit markets ground to a near-total halt (Harvey 2010).

By winter 2008 and in the wake of Lehman's collapse, panic had spread throughout the financial system and it became clear that only massive and coordinated government action would restore confidence in the financial system and thus grease the wheels of global capitalism (Harvey 2010). Minority world countries pumped trillions of dollars into the economy, banks and insurers were bailed-out or nationalised, and welfare systems across Europe and North America (initially) picked up the strain caused by huge waves of redundancies, as businesses small and large went to the wall.

What emerged as a private crisis in the world of finance quickly became (and became depicted as) a public "sovereign debt crisis" once national deficits and thus the cost of borrowing rose following the release of bailout money in 2008/9 (Watson, Brasset, and Rethel 2011). In this context, the voices of those who had identified the crisis as caused by a lack of legislative oversight, corporate malpractice, and excessive financial risk-taking, were progressively drowned out by those for whom governmental profligacy was the major issue and thus the major target for redress (Chang 2010).

In the UK, as elsewhere in Europe, a general election was fought along these lines, and in 2010, a Conservative-led government was returned promising a platform of austerity that would reign in the overblown state, restore stability to the economy, and bring Britain out of crisis. The combination of crisis-induced economic downturn and the subsequent spending cuts forming the backbone of this government's "austerity programme" quickly generated an increase in social unrest, protest, and resistance.

This crystallised in November 2010, with the first major, nationwide demonstrations – the student protests against a rise in tuition fees, the scrapping of the Educational Maintenance Allowance, and the slashing of humanities funding. Though expected to be no more than a small protest, police brutality, the deployment of the tactic of kettling,[1] and a far larger attendance than predicted saw this become the most explosive and violent demonstration Britain had witnessed for over a generation. It was echoed nationwide as coalitions formed to resist the government's austerity drive – the UK Uncut group staged media-savvy protests throughout 2010 and into 2011, while the trades union-organised "March for the Alternative" drew up to half a million people onto the streets of London in spring of that year.

Internationally, too, there were parallels of what was occurring in Britain and for many it was events beyond British shores that inspired them to engage "at home". First among these was the revolution in Tunisia and the subsequent occupation of Tahrir Square in Cairo, which formed the epicentre of the revolution that would overthrew President Mubarak in Egypt. Freeman (2012) also notes that a very similar movement was already well-established in Tel Aviv (Israel) by mid-July 2011, which was much larger than the now famous "Occupy Wall Street" (OWS) in America, but under-reported in the media internationally and hence little known.

Thus, wall-to-wall media coverage of what became known as the Arab Spring – and, in particular, attention to the tactic of occupying public space – bled into the first wave of European "occupations" those of the *Indignados* protesting their own austerity government in Spain, centred on the Puerta del Sol, Madrid.

From Madrid to New York, and in September 2011, the American OWS protests began, with thousands of activists attempting to set-up camp at the heart of the US finance industry, and ending up with a would-be permanent encampment in nearby Zucotti Park. Sister and solidarity occupations were established across North America, and by later that month a global call to action had rippled through the Internet.

On 15 October, a few hundred people gathered nervously on the steps of St Paul's Cathedral, preparing to set up London's equivalent of OWS – outside the London Stock Exchange. Though kettled and attacked by police, who were unsure of how to react to the unfolding scene, "occupiers" bedded down outside the Cathedral, were informed the following morning that the Chaplain had granted them permission to stay, and set about establishing what became the Occupy London camp that would last into the New Year and form the locus of nationwide protest. But against what?

Dominant criticisms of the Occupy movement

At its height, the Occupy London camp played host to literally thousands of protestors, ranging from the few hundred who were at any one time in residence, to the many more who came to events or followed them online via the camp's live-stream. The camp (or, rather, the individuals associated with the movement that centred around the camp) hosted some of the world's most prominent intellectuals, organised countless marches and demonstrations, published myriad newspaper articles, gave innumerable television interviews, and engaged in a long and at times difficult process of decentred, deliberative democracy determining collective positions on issues as mundane as kitchen provision to those as serious as proposals for global financial reform.

Though the movement can arguably, directly or indirectly, be credited with certain major accomplishments – notably, (1) the shift in public political discourse which has seen capitalism brought back into question after its recent "end of history" dominance (see, for example, Bunting 2011); (2) the constitutional crisis within the Church of England, which has seen the established Church take a more active and influential role in British politics, particularly vis-à-vis social justice, than has been the case for some time (demonstrated by the meetings and resignations of the St Paul's clergymen); and (3) the conscientisation and forming of a new generation of politically-engaged British citizens (i.e. people who would not have identified themselves as activists now becoming increasingly involved) – it was, throughout its lifetime, persistently dogged by one line of widely-repeated and, for many, damning criticism: that it lacked direction, that it failed to make any concrete demands, and that, ultimately, it was consequentially pointless.

Commentators left and right were quick to criticise occupiers on these grounds. In the Houses of Parliament, Conservative politicians lambasted the unwashed students sitting around talking. Publications such as the *Daily Mail* and *The Telegraph* echoed this loudly. Janet Daley (2011), writing in the latter, compared the protests unfavourably to those forming part of the Civil Rights Movement, "*which knew what they wanted*". In one widely-read article, she asked:

> "What exactly is it that would count as a successful outcome for today's protesters? They attack capitalism and want to replace it with – what? . . . So what are these people demanding? What could possibly be offered that would bring their activity to what they would regard as a satisfactory end? And who could offer it?" (Daley 2011)

Similarly, Melanie Phillips, writing in the *Mail*, complained:

> "The 'Occupy' protests are also totally incoherent . . . Nor is it clear against precisely what the tented ones are protesting. For their causes seem to multiply by the minute, including such disparate grievances as the eviction of the Dale Farm travellers, climate change and NHS reforms. A clue to the vacuousness of such protests was surely to be found in the placards brandished on Wall Street declaring: 'Occupy everything'. The words 'asinine' and 'infantile' spring irresistibly to mind." (Phillips 2011)

Even those broadly supportive of Occupy were critical. John Harris, writing in the *Guardian*, repeated a prominent critique that the movement's leaderlessness ultimately led to its being rudderless and thus slipping inevitably into oblivion. He wrote:

> "As can happen with any rudderless collection of individuals, Occupy has often seemed to turn introspective, until the issue in danger of consuming them has been the camps themselves.
>
> "Moreover, given a consensual, effectively leaderless model of decision-making – 'jazz hands', and all that – it has ended up, pretty much by definition, recurrently taking the path of least resistance. This matter of basic logic presumably explains the absence of a clever exit strategy, and why the St Paul's camp is so miserably fading away. Any alternative, no matter how creative, would always be greeted with at least some opposition, whereas staying put and fizzling out proved to be the least controversial option. On

> Occupy's terms, the result is assuredly democratic. From the outside, it also looks tragic." (Harris 2012)

Similarly Jason Hickel, himself at times heavily involved in, and undoubtedly supportive of, Occupy as a movement, had the following to say:

> "Occupy activists continue to insist that demands might alienate those who disagree with them and discourage diversity of opinions. Here we see the liberal ethic in full force ... it naively ignores the reality of political factionalism and incompatible power interests in favour of a form of inclusiveness that hamstrings any serious forward movement." (Hickel 2012)

Leaving aside the empirical fact that informal leadership structures and concrete proposals for political-economic change *did* eventually emerge from within the ranks of the London occupiers, it is our contention that these criticisms remain overly simplistic, even as they do have some merit. As we shall argue in the subsequent section, this is because, whilst imperfect and ultimately destined to dissolve, Occupy London and its camp actually represented, particularly in its early phases, an attempt at pre-figurative politics, a fact which was missed by most critical commentators.

Occupy as pre-figurative politics

> "It's very similar to the globalisation movement. You see the same criticisms in the press. It's a bunch of kids who don't know economics and only know what they're against. But there's a reason for that: it's pre-figurative, so to speak. You're creating a vision of the sort of society you want to have in miniature. And it's a way of juxtaposing yourself against these powerful, undemocratic forces you're protesting." (Graeber, quoted in Klein 2011).

"We seek to ***be the change*** *that we wish to see in society"* (Occupier).

According to political theorists from Giddens (1990) to Tilly (1997), contemporary civil political engagement and action is marked by a discontentment with, and a tendency away from, traditional political practices such as formal elections or political parties. A survey conducted at OWS, for example, showed that more than a third of occupiers declared themselves politically independent, whilst only a fifth were affiliated with, and actively involved in, political parties (Occupy Research Collective 2011). Indeed, Graeber observed that many at OWS seemed "*deeply alienated*" from current political institutions (Graber 2009, 63). This decline in formal political engagement (to paraphrase Cornwall 2008) corresponds with the rise in more fluid, undefined social movements paralleling Occupy which aim to be free of ideological or institutional constraint. Such movements have seen people taking on more nuanced attitudes towards politics with a refusal to accept all-encompassing agendas. These social movements can be understood as sustained, collective challenges against existing power structures (Tarrow 1994; Tilly 2002).

But what is pre-figurative politics? Graeber, in particular, defines pre-figurative politics as "*the mode of organisation and tactics undertaken which reflects the future society being sought by the group*" (2009, 11). The term pre-figurative politics has therefore been employed by activists and scholars alike to characterise movements that attempt to embody the values of democratic participation and power-free social engagement. Social actors within such movements are supposed to directly "*embody ... the change the movement seeks*" in all elements of social life, including movement organisation as well as identity and lifestyle

(Edelman 2001, 289), thus acting as a kind of 'working utopia' (Crossley 1999), a living model, or vision of the proposed society.

A major aspect of pre-figurative movements is that they are heavily interconnected, or networked, a term that many scholars use to explain core movement relationships. Appadurai (2000, 15), for instance, suggests that networks collaborate in a "*loose, horizontal structure*" while Juris (2008, 11) claims that they embody processes of "*de-centralisation and de-structuring*" which build "*horizontal ties*" between people based on consensus decision-making. Ideally, activists affiliate through solidarity, which transcends ideologies, "*national cultures, identities and obligations*" (Cunningham 2008, 600). Social bonds in these contexts are intended to cut across identity, location, time, and space so no one is excluded due to circumstances such as geographical location or class.

Escobar (2004, 351) endorses the concept of network relationships as a powerful ideal. He explains that these interactions are based on co-participation, in which "*negotiated views of reality may be built*". Actors are both "*receivers*" of ideas and "*potential emitters*" (Escobar 2004, 352). In Escobar's understanding, actors within networks interact with one another in total equality. He writes that the "*co-operation and collective learning*" which is part of the network process results in a democratic production of social worlds (Escobar 2004, 353). Furthermore, in Crossley's view, network relationships serve as 'proof' of the validity of the movement (1999, 830). They produce "*new plans and ideals for living*" which challenge existing social norms and values (Crossley 1999, 810). Likewise, in Deleuze and Guattari's terms, networks "*create new [0]possibilities ... invent new forms of existence, articulate alternative futures ... and create zones ... against dominating power relations*" (Deleuze and Guattari 1987, 33).

In other words, these movements represent (and are pre-figurative to the extent that they represent) spaces in which actors will follow the Ghandian mantra to "be the change you wish to see", building and living in the present structures in ways that are more decentred, less hierarchical, more democratic and ultimately more attune to solidarity than accumulation. In this sense, social movements can at times be understood as representing a kind of pre-figurative politics.

Over the course of the rest of this section, we will reflect on Occupy's attempts at governance through participative democracy, as well as specific case studies representing, despite their many drawbacks, attempts at pre-figurative politics within a contemporary social movement.

The General Assembly

> "Organization can be a model for the change one wishes to bring about." (Graeber 2009, 25)

The General Assemblies (GAs), were the principal form of governance at Occupy. Taking the form of twice-daily meetings, they provided an opportunity for occupiers to discuss various agendas and make decisions based on consensus. The goal was to produce either an "action point" (a solution), or failing that, a proposal to a set of agenda points. Described on Occupy London's website as a "*forum for peaceful assembly*" (Occupy London 2011), the GA was intended to be, "*party-less, leaderless, by the people and for the people*". GAs were founded on the principle that "*people are not perfect ... one person alone even less so!*" and were organised in opposition to "*top-down, macho, heroic models of leadership*". By implicitly rejecting current constitutional models which "*destructively concentrate power among a few key individuals*", they were also an attempt to involve all members in the political sphere, regardless of age, status, gender, or intellectual ability.

At most GAs, occupiers would sit close together. Ideally, all members of the Occupy encampment were supposed to participate, so the meetings provided a rare opportunity for everyone to gather together in one place. They were coordinated by an elected facilitator whose function was to ensure that each individual voice was heard and each agenda point properly discussed. Facilitators would also gauge the assembly's reaction to the proposed agenda by taking "temperature checks" on it, which involved participants making hand signals to demonstrate agreement, or a desire to contribute further to the discussion. A key purpose of the GA was to encourage interaction through dialogue and face-to-face communication, something many occupiers regarded as essential for a democratic society. For some, it seemed that face-to-face communication was preferable to other forms of communication such as cyber interaction or written documentation, based on the notion that pure democracy meant not defining yourself. Speech and discussion were seen as organic and fluid, whereas written statements were seen by some as rigid and exclusive as they might not allow for diverse perspectives. As one informant told me, true democracy comes from "*the heart*" instead of pieces of paper. It is also of course significant that written statements can alienate where literacy is low.

In essence, the GAs were an attempt to put the *process/means* of democracy ahead of its *end*. Rejecting the consequentialist position that underpins constitutional democracy in its dominant liberal format, GA processes posited an alternative in which the end of getting things done or taking a decision is not privileged over the process of not alienating, steam-rolling, or excluding those upon whom such decisions may impact. In this structure, individuals represented themselves, and all participated in those GAs in which they felt they had an interest. The GA was thus clearly an attempt not only at governance, but at shining a light on the inner contradictions of constitutional, representative government such as it exists in British society.

The Tent City University

Tent City University (TCU) was one of the very first branches set up as part of the Occupy London movement. Neil chaired the inaugural working group meeting and subsequently played a central role in organising the University's activities until these and the camp itself wound down towards Christmas. Here, he reflects on the University's work and modus operandi, and will present three examples which indicate its tendency towards a form of pre-figurative politics.

The first of these can be seen in TCU's broad approach to learning. Though many TCU organisers came from formal academic backgrounds, all were of the opinion that the structure of contemporary academia can and does reproduce both the injustices of the dominant neoliberal political-economic system and the class divisions and inequality on which this rests. As Paulo Freire (2007) argued, "*there is no such thing as a neutral education process*", and in the context of the contemporary marketisation of higher education, we felt that formal universities functioned less as creators of critical, free-thinking human beings and more as a production chain for the integration of the young into the market economy. Related to this, we believed that the myth of equal access to education and the process of grade-standardisation had led university qualifications to become a pillar of social hierarchisation (see Bourdieu 1998 for a fuller discussion of this). In today's marketplace, of course, those with a Masters are deemed to be worth more than those without a degree, with the accumulation of qualifications thus constituting a path to the accumulation of status, wealth, and power (Bourdieu 1998).

TCU openly rejected this. Our motto was "anyone can teach, everyone can learn" and we sought to promote an approach to learning that prioritised *process over end-point* and valued the skills all people have to share and the capacity all people have to learn. What did this mean? It meant that our workshops were given by bicycle mechanics and electricians as well as by academics, that every speaker sat non-hierarchichally among the crowd and that all were welcome to participate in any way desired. It also meant accessing perspectives from across the political spectrum and creating a space in which dialogue – rather than the top-down imparting of information – was the norm. Finally, it meant avoiding conferring qualifications or certificates demonstrating that an individual had learnt *x* or *y* or acquired any particular, categorised qualification.

Our second major challenge was to break down some of the communication barriers that we felt were constructed by the dominant political-economic and social status quo. While research increasingly reveals that humans are hard-wired to be empathetic (Rifkin 2010), recent neurobiological studies have demonstrated that empathy is reduced by lack of interaction and mobility across social groups (Shamay-Tsoory 2011). In our society, we felt these divisions to be especially acute and we believed this to be a significant contributing factor to prevailing injustice and inequality. In order to challenge this, we attempted to create an open, free, and welcoming space in which people from all walks of life could come together. By having ex-bankers share stories of the wrongs of their industry, by having migrant workers tell of the abuses they faced while cleaning the buildings home to that industry, and by facilitating debates between those from radically different political and social spheres, we strove to bridge divides and contest the many forms of division, privilege, and discrimination that separate us from each other. Evidently, we were aware that what we created was fleeting and of small scale – the goal however was to prefigure a context in which new collectivities, communities, and identities could (at least begin to) emerge.

Our third challenge regarded the use of public space. Though, intuitively, British citizens may assume they have the right to peacefully assemble at the heart of London to voice grievances and air concerns, in reality they do not. As in city centres across the country, many of London's nominally public squares and thoroughfares have been parcelled off to private landowners or corporations who have the right to legally exclude anyone from rough sleepers to protesting students (Minton 2012). Again, we rejected this as well as the neoliberal, anti-commons philosophy underpinning it. In order to manifest that rejection, we used TCU to conduct a series of "flash teach-outs" – a combination of education and direct action, where we assembled unexpected and en masse to stage public lectures and host open debates. After the success of our teach-outs in front of the Bank of England, the Corporation of London fenced off what most previously thought the public square outside it. When asked why they had done so, they told us it was to "prevent disruption". When we spread this information, floods of London citizens complained, armed now with the awareness previously denied them.

The Welfare and Well-being Working Group

> "Occupy is curative ... we are the antibodies to the disease." (Occupier)

This section explores the activities of the Welfare and Well-being Working group (hereafter, WWG) which was set up about one month into Occupy London. The group initially formed in response to a "call out" claiming that the camp had become a "beacon for those in need". In particular, volunteers were needed to help with overwhelming numbers of people experiencing a range of problems such as activist burn-out, alcoholism, schizophrenia, aggressive tendencies, and so on. Indeed, because of Occupy's policy of inclusivity and its welcoming stance, it attracted many people who were drawn to the camp for reasons other than activism, such as the support, food, and shelter freely offered by occupiers.

The WWG was run by volunteers who took turns staffing the "welfare tent". Decorating the space with cushions, lamps, and sofas, volunteers intended to create a calming and supportive environment in which occupiers and others could drop in to rest, talk, and relieve the stresses of day-to-day activism and life at the camp. People could walk into the tent without appointment and could stay for as long as they liked. Volunteers did their best to provide services without time limits and constraints, which was, in their view, explicitly and intentionally in contrast to the compartmentalised and privatised nature of mainstream therapy today.

The WWG had some clear, political motives. As one volunteer noted, "*we have to welcome all kinds of people into Occupy, especially those who have been rejected by society*". By this, she meant those large numbers of the homeless or mentally ill who were becoming increasingly drawn to Occupy London.

The WWG eventually became a venue where people converged to discuss alternative approaches to mental health. Volunteers and users of the welfare tent unanimously agreed that popular understandings and treatment of mental health had been corrupted by capitalist ideology. Many occupiers, for example, attributed mental health problems such as schizophrenia to capitalist values and lifestyles rather than biological or physiological factors. Poor mental health was often blamed on the market-driven values of capitalism, specifically to the prevailing "*atmosphere of fear, aggression and competition*" which is unsustainable for our "*psyches, souls or communities*" (Altman et al. 2012).

Moreover, some WWG members argued that the mainstream healthcare system does not improve these problems, but rather perpetuates them, as it is governed by the need to fulfil governmental quotas, and to create stable, functioning individuals whose sole function is to be able to work and contribute economically to society. In accordance with this view, some volunteers argued that drug treatments which are so widely used among the NHS, are used to supress individual creativity and difference, whilst one-to-one therapy creates individualised, privatised relationships based on economic exchange rather than a real care motive. Overall, they argued that we should scrutinize the current socio-political environment which is unhealthy for our psyches.

People involved in the welfare campaigns came from very diverse backgrounds and WWG members eventually came to include shamans, psychotherapists, psychiatrists, Reiki healers, social workers, and counsellors alongside medical service users and mental health activists. Weekly WWG meetings were often made up of volunteers and, at times, the people they strove to help. Some volunteers had no training or expertise, but maintained that life experience and human compassion were the basic and vital skills needed to help others. There were no formal rules or guidelines for how to treat welfare users, so people's practice varied extensively, according to personal preference or professional paradigms. Some therapists would sit inside the tent and wait for patients to enter as and when they chose to, while others preferred to stand outside chatting informally to people, taking them into the tent only if they wanted to talk in greater depth. These volunteers preferred a much more informal, spontaneous, and user-led approach to healing, being deeply critical of conventional therapeutic practice because it was perceived as embodying certain characteristics that for typify capitalist society, particularly the power dynamics and hierarchical structure.

Others, however, disagreed with this loose informal approach to treatment, and argued that in therapy, structure is important for empowering patients. These volunteers often encouraged patients/users to visit the tent on a regular basis, as in more conventional therapeutic settings. A few volunteers also drew on outside support from pre-existing services. Some frequently liaised with the charity Broadway, for instance, who aim to house and support houseless people. Others used referral systems and sectioning in cases where they felt they did not have the sufficient resources to deal with a patient. These particular interventions were extremely controversial as Occupy aimed to be autonomous and self-sufficient, hence many bemoaned such decisions, arguing that external agencies such as Broadway are entrenched in the capitalist, statist system and are thus unreliable and oppressive. This issue often led to internal debate and contention among the WWG.

Despite some internal disagreement, the WWG more often than not reflected Occupy's wider philosophy of living the example they wished to see in society. Members who opposed the rationalisation of mental illness within institutions drew on their desire to provide a space which was inclusive of everyone and thus created a healing environment in which people interacted spontaneously and freely, unencumbered by status or hierarchy. Such members set up alternative healing treatments among the community which were often based on holistic therapies. "Encircling therapy" (sometimes also called a "containing circle") was how some dealt with a person experiencing a tantrum or psychotic episode. On seeing someone in distress, familiar people and friends were encouraged to form a circle around the distressed person and hold (contain) them (*without* physical contact) in an attempt to calm them and make them feel supported, and also to ensure he/she did not hurt him/herself or others. The person may then be taken aside and encouraged to talk about their feelings and problems. I suggest there are a number of things at play in the act of encircling. Symbolically, it represents a mode of healing in which power/hierarchy is decentralised because all of those standing in the circle play an equal part in the healing process. Encircling, I was told, relies on the voluntary actions of people who happen to be present at the time, this being in stark contrast to the formal patient/professional relationship. Thus, encircling both structurally and symbolically represented horizontal politics in a kind of embodiment of the type of community occupiers wished to become.

In line with Occupy London's self-sustainable DIY (do it yourself) approach, skill-sharing workshops were also a popular way of encouraging everyone to be involved in the process of healing. Those focussed on mental health included "co-counselling" and "how to get someone through a crisis". The workshops were available to everyone, on the grounds that anyone can *do* counselling and extensive training is not necessary. People who attended such a workshop were encouraged to pass on their knowledge to others. I witnessed many instances of what could be called co-counselling, in which actors took on the dual role of carer and cared-for, taking turns to ask each other questions and to discuss problems and feelings. Discussion groups focussed on mental health also encouraged people to "speak out" (a common trope in mental health activism), sharing stories and past experiences which often included accounts of trauma or mental illness. The purpose of these talking groups was diverse. Individuals were encouraged to explain their own personal and unique story to increase awareness of mental illness and the stigmatisation it incurs, whilst speaking out demonstrated the importance of both speaking and listening in communal settings.

In sum, people aimed to build a community in which care was "*a part of life, not a job or occupation*". As one welfare member exclaimed, "*paying people to care undermines the human aspect of communication*"; she longed for people to have the desire to listen to her, instead of being paid to. Although the WWG encompassed very diverse members, people united over their concern for the well-being of others and their belief that the current British care system is lacking in human concern and compassion.

Discussion and conclusion

It is important to begin this discussion with an acknowledgement of the fact that Occupy London had many problems and that, for all its valiant efforts, its striving for pre-figuration was often unsuccessful, which necessarily raises questions about whether pre-figuration is a viable political strategy.

In the first instance, despite their attempts, the GAs were often highly fractious events, which sometimes came to be dominated by particular individuals or factions. In this sense, they at times failed in their own terms, since not all were able (or willing) to participate equally under such circumstances. Moreover, not everyone who was implicated in or impacted by a decision was able to attend all relevant GAs, such that the very purpose of the GA as a hyper-democratic decision-making forum was undermined.

This fact, coupled with the reality that GAs at times meandered frustratingly through minor details toward non-conclusions, suggests that the GA, while useful as a challenge to the increasing vacuity of contemporary representative democracy, lacked certain of the practical advantages of representation and delegation. To judge the GA using the consequentialist lens of the liberal democratic, representational structure of contemporary British governance may be unfair, but so would denying the benefits that representation and delegation can bring.

The WWG also faced problems. Indeed, in line with the major criticisms examined above, it often struggled with structurelesness and leadership. Working groups were supposed to emulate Occupy's democratic decision making processes, but some found it difficult to work in an environment in which relationships were not defined by status, wealth, or power, specifically in the pseudo-medical context we inhabited. Many were used to drawing on their formal training experience or expertise to situate themselves and relate to others in their line of work. As such, a tacit hierarchy and system of leadership did actually emerge, whereby leaders were self-elected, based on their level of confidence, on recognition, or the length of their involvement in the WWG. These leadership roles were often appropriated by those more inclined to clinical and professional roles, meaning that at times the WWG came to reflect precisely those dominant structures it sought to challenge. Though perhaps a failure in pre-figurative terms, this might not necessarily have been a negative outcome, since some of those we cared for truly needed expert assistance.

Parallels exist here also with TCU. Though we endeavoured to embody a kind of pre-figurative horizontalism, the positionalities of those involved in driving the University forward necessarily led to the reproduction of certain dominant social hierarchies. TCUs central group of organisers were all either current or former university graduates, many of whom had traditional teaching experience, and to whom other organisers often deferred. Similarly, though myriad efforts were made to include speakers and sessions from beyond the academy, the majority of TCU speakers (and all of the best attended sessions) were (those conducted by) established academics or respected social thinkers.

Moreover, in practice, we found hierarchy and unofficial leadership/delegation to be necessary. This is not to diminish the real attempts at mutual respect and, at least, a minimisation of hierarchic authority; it is merely to acknowledge that organising complex events across time and space necessitated the kind of division of labour and delegation that does not sit perfectly with an absolute pre-figurative horizontalism.

Perhaps what these drawbacks highlight is that pre-figuration is not an easy goal to accomplish. Living the future in the present is an admirable goal, but sometimes drawing on the structures of the past can help in that transition. If nothing else, the very *attempt* to pre-figure, while not invalidating all that dominant contemporary structures stand for, at least highlights some of their failings, some of the ways in which they do not live up, in practice, to the ideals upon which they are putatively built.

In this, it is worth noting the many tangible, democratic, egalitarian, and thus pre-figurative successes the movement did enjoy. The GAs and the WWG, for instance, provided a space in which certain individuals could become empowered with strong feelings of autonomy, especially in cases of those who are typically excluded and marginalised by society such as the homeless and mentally ill. The deconstruction of hierarchy in treatment particularly allowed this.

Similarly, with TCU, while key organisers and speakers were deeply implicated in the neoliberal academy and thus complicit with many of its injustices, the very location and self-image promoted by TCU allowed large numbers of people not involved in formal education to engage in and benefit from the kinds of learning spaces normally only provided in formal educational contexts. It is of real significance that so many self-described "ordinary" citizens were able to access the kinds of critical thinking/learning normally reserved for the relatively very privileged.

Crucially, we must recognise that an important aspect of any pre-figurative political action is the recognition that a protest, as with a social formation, is an ongoing journey as opposed to a pre-defined destination. Pre-figurative politics are thus processual, meaning that even when one form evaporates or fails to achieve its goals, its essence may continue in another, with lessons incorporated "from the road". This is evidently the case with Occupy. While the significance of the camp's physical presence cannot be underestimated, and while the disbanding has undeniably effected the nature of contemporary political action in the UK, the ethos and engagement embodied by those (blooded) at Occupy still lives on. This can be illustrated in myriad ways. There are, for instance, a variety of Occupy-related collectives which continue to organise (Occupy Research or Oral History providing two examples). In the case of TCU, those at the heart of its organisation have drawn on its models to build a nationwide network of popular political-economic education (see www.peliteracy.org). While those drawn into its network now spearhead for a such as the Coalition for Resistance or UK Uncut. In this, lessons learnt at St Paul's will apply.

Notes

1. Kettling is a crowd control tactic deployed by British police in large demonstrations. It involves confining protestors within a limited area. It has been described as arbitrary detention, but has been upheld as a lawful police tactic by the European Court of Human Rights
2. Quotes taken from General assemblies at Occupy London during 2011–12
3. Much of this section draws on an analysis offered previously, in *The Independent*. See here: http://www.independent.co.uk/opinion/commentators/every-day-more-gather-eyes-are-opening-and-the-shackles-of-apathy-loosening-6256934.html

References

Altman, S., J. Bossewitch, E. Stiends, B. Glaser, S. Harper, A. Imai, R. Liebert, and M. Osborn. 2012. *Mindful Occupation; Rising Up without Burning Out*. Occupy America: Published under the Creative Commons Licence. Accessed January 18, 2012. http://mindfuloccupation.org/files/booklet/mindful_occupation_singles_latest.pdf

Anonymous. 2012. "Talking about a Revolution." *The Economist*. Accessed October 12, 2011. http://www.economist.com/node/21552179

Appadurai, A. 2000. "Grassroots Globalization and the Research Imagination." In *Globalization*, edited by A. Appadurai, 1–21. Durham: Duke University Press.

Bourdieu, P. 1998. *State Nobility: Elite Schools in the Field of Power*. London: Polity.

Bunting, M. 2011. "Occupy London is a Nursery for the Mind." *The Gaurdian*. Accessed January 19, 2013. http://www.guardian.co.uk/commentisfree/2011/oct/30/occupy-london-nursery-mind

Chang, H-J. 2010. *23 Things They Don't Tell You About Capitalism*. London: Allen Lane.

Crossley, N. 1999. "Working Utopias and Social Movements: An Investigation from Radical Mental Health Movements in Britain." *Sociology* 33 (4): 809–30.

Cunningham, H. 2008. "The Ethnography of Transnational Social Activism: understanding the Global as Local practice." *American Ethnologist* 26 (3): 583–604.

Daley, J. 2011. "The Occupy Movement Has Failed the Essential Test of Protest." *The Telegraph.* Accessed September 3, 2012. http://www.telegraph.co.uk/finance/financialcrisis/8901300/The-Occupymovement-has-failed-the-essential-test-of-protest.html

Deleuze, G., and F. Guattari. 1987. *A Thousand Plateaus: Capitalism and Schizophrenia.* London: Athlone Press.

Edelman, M. 2001. "Social Movements: Changing Paradigms and Forms of Politics." *Annual Review of Anthropology* (30): 285–317.

Escobar, A. 2004. "Other Worlds are (Already) Possible: Self-organisation, Complexity and Post-Capitalistic Cultures." In *World Social Forum: Challenging Empires*, edited by J. Sen, and P. Waterman, 349–358. New Delhi: The Viveka Foundation.

Freeman, D. 2012. "The Invisible Hand has Given me the Finger: Neoliberalism, Protests and Contested Moral Economy in Israel." Paper presented at the UCL Social Anthropology department seminar, London, May 2.

Freire, P. 2007. *Pedagogy of the Oppressed: 30th Anniversary Edition.* New York: Continuum.

Giddens, A. 1990. *The Consequences of Modernity.* Cambridge: Polity.

Graeber, D. 2009. *Direct Action: An Ethnography.* Oakland: AK Press.

Harris, J. 2012. "Occupy London: 'What Went Wrong?" the *Guardian.* Accessed August 15. http://www.guardian.co.uk/commentisfree/2012/feb/13/occupy-london-what-wentwrong

Harvey, D. 2010. *The Enigma of Capital and The Crises of Capitalism.* London: Profile Books.

Hickel, J. 2012. "Liberalism and the Politics of Occupy Wall Street." *Anthropology of this Century.* Accessed December 3. http://aotcpress.com/articles/liberalism-politics-occupy-wall-street/

Juris, J. 2008. *Networking Futures: The Movements against Corporate Globalisation.* Durham: Duke University Press.

Minton, A. 2012. *Ground Control: Fear and Happiness in the 21st Century.* London: Penguin.

Occupy London. 2011. "Initial Statement. Accessed May 12, 2012. http://occupylondon.org.uk/about/statements/initial-statement

Occupy Research Collective. 2011. "Preliminary Findings: Occupy Research Demographic and Political Participation Survey. Accessed April 26, 2013. http://www.occupyresearch.net/category/survey/

Phillips, M. 2011. "Disillusionment with the Entire Political Class is Behind the St Paul's Protest." *The Daily Mail.* Accessed December 7, 2011. http://www.dailymail.co.uk/debate/article-2052650/Occupy-London-Stock-Exchange-Disillusionment-political-classblame.html

Rifkin, J. 2010. *The Empathic Civilization: The Race to Global Consciousness in a World in Crisis.* New York: Jeremy P. Tarcher.

Shamay-Tsoory, S. G. 2011. "The Neural Bases for Empathy." *Neuroscientist* 17 (1): 18–24.

Tarrow, S. 1994. *Power in Movement: Collective Action, Social Movements and Politics.* Cambridge: Cambridge University Press.

Tilly, C. 1997. *Roads From Past to Future.* Lanham, MD: Rowman & Littlefield.

Tilly, C. 2002. *Stories, Identities, and Political Change.* Lanham, MD: Rowman & Littlefield.

Watson, M., J. Brasset, and L. Rethel. 2010. "The Political Economy of the Sub-Prime Crisis: The Economics, Politics and Ethics of Response." *New Political Economy* 15 (1): 17.

The authors

Neil Howard is currently completing an MPhil in Development at the University of Oxford. He was an activist at Occupy and played an inaugural role in setting up the Tent City University.

Keira Pratt-Boyden (corresponding author) is a recent Masters graduate of Anthropology at University College London. She spent three months engaging in research at Occupy London. Much of her contribution to this article draws on her thesis.

Dutch civil society at crossroads

Rik Habraken, Lucas Meijs, Lau Schulpen, and Cristien Temmink

The Netherlands have a rather large civil society. Interestingly, Dutch politicians and civil society organisations do not recognise this themselves, voicing frequent appeals for more civil society. This can be explained by the typical development of Dutch civil society through a system of pillarisation, based upon concepts of subsidiarity and sovereignty in one's own constituency, leading to a civil society that at the end has outsourced to the government important functions like financing, governing, and legitimising. Two cases (traditional and contemporary) are presented to show the consequences. Finally, lessons for other countries at civil society crossroads are drawn.

La société civile néerlandaise à la croisée des chemins

Les Pays-Bas ont une société civile assez importante. Il est intéressant de constater que les personnalités politiques et les organisations de la société civile néerlandaises ne reconnaissent pas ce fait elles-mêmes ; en effet, elles lancent des appels fréquents à un élargissement de la société civile. Cela peut s'expliquer par le développement de la société civile néerlandaise à travers un système de polarisation, basé sur des concepts de subsidiarité et de souveraineté dans son propre groupe constituant, ce qui engendre une société civile qui, au final, a sous-traité au gouvernement des fonctions importantes comme le financement, le gouvernement et la légitimation. Deux cas (traditionnel et contemporain) sont présentés pour illustrer les conséquences. Enfin, des enseignements à l'intention d'autres pays dont la société civile se trouve à un carrefour sont proposés.

La sociedad civil holandesa en la encrucijada

Holanda tiene una sociedad civil bastante grande. A pesar de ello, es interesante ver que los políticos y las organizaciones de la sociedad civil de Holanda no se dan por enterados, ya que a menudo hacen llamados a que la sociedad civil tenga mayor presencia. Lo anterior puede tener su explicación en el desarrollo de la sociedad civil holandesa, el cual se produjo típicamente, a través de un sistema de polarización basado en los conceptos de subsidiariedad y de soberanía a nivel de base, que finalmente dio lugar a una sociedad civil que subcontrató al gobierno para implementar ciertas actividades importantes, por ejemplo, el financiamiento, el acto de gobernar y la legitimización. Para mostrar las consecuencias de ello, se presentan dos casos, uno tradicional y otro contemporáneo. Por último, se examinan aquellos aprendizajes que pueden ser relevantes para las sociedades civiles de otros países que se encuentren en la encrucijada.

Sociedade civil holandesa na encruzilhada

Os Países Baixos têm uma ampla sociedade civil. Interessantemente, os próprios políticos e organizações da sociedade civil de lá não reconhecem isto, fazendo frequentes apelos para haver mais participação da sociedade civil. Isto pode ser explicado pelo típico desenvolvimento da sociedade civil holandesa através de um sistema de polarização, baseado em conceitos de subsidiaridade e soberania em seu próprio eleitorado, levando a uma sociedade civil que, no final, tem delegado ao governo funções importantes como financiar, governar e legitimar. Dois casos (tradicional e contemporâneo) são apresentados para mostrar as consequências. Por fim, são extraídas lições para outros países cuja sociedade civil se encontre em uma encruzilhada.

Introduction

Even after ten years Kendall and Anheier's (2001, 228) quote remains valid:

> "In some countries, policy makers and many researchers tend to see NPOs [Non-Profit Organisations] less as a 'sector' as such but rather as 'discrete' actors in 'vertical' fields like education, social services, training and employment, or the environment, the Netherlands being the clearest example."

Perhaps due to these vertical fields of activity, the Dutch civil society is of considerable size and importance (Burger and Dekker 1998). In fact, the Netherlands is one of the highest scoring nations on the Civil Society Index (Salamon et al. 2004, 78) with clearly a rather large civil society (CIVICUS/De Nieuwe Dialoog 2006).

Although this is true from an international, academic perspective, it is certainly not reflected in current Dutch politics and public perception and discourse. The general opinion in political circles, for instance, seems to be that civil society is lacking and needs to be introduced in the Netherlands. It is one of the underlying goals for the 2007 Social Support Act (*Wet Maatschappelijke Ondersteuning*, WMO). The WMO is part of a:

> "broader policy, emphasising individual responsibility in health care, both at the insurance side as well as the provision of care side ... The aim of the Social Support Act is participation of all citizens to all facets of the society, whether or not with help from friends, family or acquaintances; the perspective is a coherent policy in the field of the social support and related areas."[1]

In reaction, citizens, large civil society organisations, and politicians wonder what this concept of civil society is about. In many cases civil society seems to be limited to citizens helping themselves and to employing volunteers as part of the service. Large civil society service providers are afraid that the call for privatisation and tenders will bring in market organisations, not recognising their own history as private initiatives.

The main explanation for this reaction seems to be that the Dutch civil society *sector* has lost its links with civil society. As Stubbe (2006) remarks, civil society sectors differ in their structural "non-profitness". The five structural conditions as defined by Johns Hopkins' Comparative Non-profit Sector Project (organised, private, non-profit-distributing, self-governing, and voluntary) vary in degree and examples may qualify more easily on one criterion than another, leading to different degrees of non-profitness. For example, the less private and self-governing a civil society organisation is (e.g. more part of or controlled by government), the more it moves towards the public, governmental sector. The extent to which the organisation distributes its profits to members moves it towards the private, business sector. Hupe and Meijs (2000) show that Dutch public organisations are so hybrid that the theoretical differences between civil society and government organisations in, for example, primary education and social housing have very little impact on their functioning. Primary education schools differ from each other because they are in different neighbourhoods, not because one is governmental and the other is civil society. In practice, the rather large Dutch civil society sector is very limited in its self-governing and voluntary input and is often neither perceived as a separate sector nor as a separate part of society next to government.

Due to the specific Dutch context, which will be described later, Dutch non-profit and civil society organisations came to regard the state as their primary source of financial support, and as a consequence also as their source of legitimacy instead of their constituencies (RMO 2010). In fact, they became idle about developing alternative sources of support, both emotionally and financially. Reliance on state funding led to extreme forms of resource dependency and associated goal displacement (Pfeffer and Salancik 1978) – with national organisational goals reflecting those activities eligible for funding rather than the mission of the organisation.

The search for reconnecting to civil society, if only for legitimacy reasons, is taken here as the most important contemporary crossroad for the development of Dutch civil society. In both this reconnection process and the proceeding delinking process, the relation between the state and civil society organisations is central. To grasp the historical roots of this relationship, this article starts by sketching five phases in the development of Dutch civil society. It then provides two cases illustrating the crossroad. The case of the recently established Makassar neighbourhood community in Amsterdam reflects the ideas of the last phase and is thus exemplary of the felt need to create civil society and for citizens to take responsibility. Being successful, it has now the possibility to apply for governmental subsidy, running the risk of becoming a traditional, governmentally-controlled civil society organisation. The Dutch Council for Refugees (VWN, *Vluchtelingenwerk Nederland*) – our second case – is a typical example of the start, rise, and uncertain future of many contemporary Dutch CSOs. Moreover, it is a clear case in which the state was not only instrumental in setting up the organisation in the first place, but also in funding service delivery activities, creating tensions with the advocacy role of VWN. We end with a brief discussion about the lessons that can be learnt from the Dutch experiences for the development of civil society in other countries.

Five phases of development

Hupe and Meijs (2000) sketch the development of civil society in the Netherlands on the basis of four phases: (1) pre-pillarisation: an emancipation process; (2) pillarisation: serving the public by serving your own group; (3) de-pillarisation: going public; and (4) going private: introducing the market. Since then, a fifth phase can be distinguished which is labelled "*going for ownership: reintroducing civil society*" (Meijs 2004). This last phase started around 2003 and the Social Support Act of 2007 can be seen as codifying this fifth phase in the domain of social welfare. It must be understood that these five phases are in general terms valid for the whole civil society but do insufficient justice to the diversity within that civil society. The phases are, for instance, more visible in large paid-staff service providers than in small all-volunteer groups or in cultural institutions. In fact, for many nationally operating civil society organisations, the five phases have real consequences for the paid staff at headquarters and limited consequences for the local chapters (see Meijs 2007). Nevertheless, this Dutch development shows that CSOs face a complex environment in which objective developments, subjective political framing, and general institutional economic and political developments are intertwined and interact.

Phase 1. Pre-pillarisation: an emancipation process

As a result of the 1848 Constitution, the Dutch educational system was almost completely regulated by the national government and a national ministry for education (see Hupe and Meijs 2000, 42). While this was what the ruling Liberals were aiming for, there was great tension with religious and class-based groupings that had different expectations of the role of the state, particularly in their aspirations for the educational system. Denominational groupings, in particular Roman Catholics, wanted religion-based education to be supported by the government. For the working class, the dominant debate was about social issues and social recognition. This struggle for emancipation from different groups aiming for support by the government for their own organisations, and with the so-called education battle (*schoolstrijd*) as an important issue, ended more or less in 1917. At that time, the Socialist party and the Liberals traded their goal of general voting rights in exchange for the objective of Roman Catholic and Protestant groupings for financial support from the government for religious schools. This can be seen as a starting point of the Dutch state-funded civil society as it opened the door to large-scale subsidising of religious and class-based organisations in almost every social welfare and non-profit field. Through this funding of CSOs tensions between different groups in society were pacified.

Phase 2. Pillarisation: serving the public by serving your own group

The process of subsidising multiple religious and class-based organisations led to the phenomenon of pillarisation. Dutch society was typified as a peculiar kind of plural society with discrete "vertical" segments that have their own separate social and political organisations. Unlike a society "horizontally" stratified by class, these segments have a vertical character with the base layers and the elite within each segment being related to each other. These segments that made up Dutch society are referred to as pillars (*zuilen*) and the process of segmentation as pillarisation (*verzuiling*). Pillars are "*blocs of institutions and members, delimited from each other in different sectors of society along the same ideological demarcation lines*" (Pennings 1991, 1). Pillarisation can then be understood as diversification combined with consensus-building at the top where the elites of the different segments meet each other and construct and maintain a basic consensus. Political majorities can only be formed through coalitions (Hupe 1993; Hupe and Meijs 2000). The process of pillarisation in society created a policy framework for the development of voluntary civil society service delivery organisations, mutual support associations, and campaigning organisations. It remained the dominant model through the twentieth century into the post-World War Two period. Government just funded the paid staff and the headquarters. The concept of pacification led to an almost automatic system of treating and funding all pillars in the same way: all pillars got their fair share out of the national taxes.

Phase 3. Depillarisation: going public

As a result of post-war modernisation, pillarisation as a form of social control of individual citizens was undermined. Bax (1988) identified three modernising forces that contributed to the decline of pillarisation. The first was the rise of the welfare state in which the standard of living rose and social legislation emerged. This allowed individuals to become less dependent on their pillar, while the pillars themselves became more dependent on the state. Second, with the welfare state and post-war technological innovation came increasing mobility (geographic and social), rising levels of education and new windows to the world with television. Third, secularisation meant that the impact of denominational sets of norms and values (ideologies) diminished (also see Hupe 1993, 376–7). Hupe and Meijs (2000, 151) state that this process of depillarisation created the basis for what the Dutch call the "*calculating citizen*": an active citizen with very few, loose ties to a traditional pillar, who takes for granted the benefits of the welfare state, knows his or her rights, and opportunistically follows rules only if this provides benefits. In practice, it created a divide between pillars looking to government for funding and their former constituencies who were getting services from different pillars.

However, the pillared structure of civil society (e.g. in primary education) to a large extent remained intact. In this context, Van Doorn (1984, 41) states that what has been called depillarisation "*to an important degree was the redefinition and regrouping of confessional interests*". Over the years this regrouping, visible in (forced) mergers, led to the formation of larger civil society organisations which either served a much larger geographical area or moved from representing one religious grouping to encompassing many. Through these mergers, civil society organisations more or less adapted to field-wide norms (Powell and Friedkin 1987). These changes led Hupe and Meijs (2000) to conclude that by the end of the third phase, the state and the civil society were very much intertwined. According to Dekker (1998, 127):

> "it is becoming increasingly difficult to distinguish the traditional private non-profit organisations from the many independent public bodies, PGOs (para-government organisations) and quangos (quasi-autonomous non-government organisations) that have been established by government in recent decades."

Phase 4. Going private: introducing the market

In the fourth stage, two forces impact the structural configuration of organisations in civil society. The bureaucratic control of subsidised civil society was becoming a burden to the constituent organisations and government bodies providing subsidies (Hupe and Meijs 2000). At the same time, privatisation and liberalisation – just as in the UK, US, and elsewhere – became buzzwords in public policy. New policies forced large service-delivering organisations to work under conditions of market powers. The existing (geographical) monopolies held by specific organisations were brought to an end. Perhaps the most significant change was the government policy of giving clients a budget that they could spend on activities of certified service-delivering organisations, instead of financing these organisations directly. As clients could then choose which organisation they would contract, civil society moved inexorably towards the market sector. They encountered competition, professionalised, and responded to pressures to work more efficiently and effectively and to meet quality standards. Without these changes, they could neither qualify for state subsidies nor compete with other service providers – non-profit or for profit.

Phase 5. Going for ownership: reinventing civil society?

The start of the new millennium has been turbulent for the Netherlands: an economic recession in 2001, a political assassination in 2002 and, for the first time in years, a budget deficit in 2003. The state's response included a re-evaluation of its subsidising policy. This not only influenced the large all paid-staff service delivery organisations, but also the paid-staff headquarters of larger nationwide volunteer organisations. The Ministry of Healthcare, Welfare, and Sport's subsidies to 254 civil society volunteer organisations were proposed to be cut back or even removed (Meijs 2004). Also in other fields the same process took place. Organisations in development cooperation, for instance, saw a combination of stricter subsidy rules based on competition and a reduced budget (Schulpen 2006, 2009). It is extremely likely that this combination of less (certain) government funding and strict rules for funding eligibility and spending guidelines will be even further strengthened in coming years.

The response of affected organisations has been twofold: to search for other sources of state support and to attempt to realign themselves as civil society organisations that can draw support – financial and otherwise – from within civil society. Most did find ways of securing new forms of state funding but it has not proved to be as significant as before. After the 2007 Special Support Act, for most social services and welfare organisations reconnecting to or making use of civil society became the new buzzword.

Makassar neighbourhood community

The fifth phase in the development of Dutch civil society thus not only led to drastic changes in the governmental subsidy system, but also to a revisiting and reshaping of the civil society discourse. Emphasis gradually shifted to the resourcefulness of citizens to solve their own issues and depend less on the state for services and subsidies. Civil society's role is then considered in empowering citizens to take that responsibility, and in delivering services that the state cannot – or will no longer – provide. Hence, two functions of civil society are emphasised: service delivery and community (social capital) building. Keywords that resonate are social capital, social cohesion, and active citizenship. Central to the discourse is the idea of changing roles and responsibilities between government, civil society, and citizens. However, it is not clear yet where responsibilities and roles of each start and end. Therefore, some argue that government priorities have been "*motivated by a strong desire to move traditional public responsibilities either to civil society or to citizens themselves*" (CIVICUS/ De Nieuwe Dialoog 2006, 41).

The Makassar community in the Indonesian neighbourhood of Amsterdam is a clear example of this fifth phase where government policies intermingle with the reintroduced notions of civil society. In 2007, a Dutch government policy priority area was the so-called neighbourhood approach *(wijkaanpak)* to improving liveability in severely deprived neighbourhoods. The Indian Indonesian neighbourhood of Amsterdam-East was declared a high priority area in the framework of this policy approach. Originally built in the early 1920s for housing harbour labourers, the area is divided into four quadrants: Ambon square, Timor square, Sumatra Park, and Makassar square. With the decline and finally disappearance of port functions in the 1960s, the neighbourhood remained isolated and decayed. From the 1970s onwards, immigrant workers and their families settled in the neighbourhood. Currently the area has over 23,000 young people from diverse ethnic backgrounds. Dutch natives are a minority; 65 per cent of the population consists of immigrants with a predominance of groups from Morocco, Surinam, Turkey, and the Dutch Antilles. An estimated 100 different languages are being spoken (Sterk 2011). Social-economic issues at play are high unemployment, low incomes (with 25 per cent of households below the poverty line), high school dropout, social isolation (in particular of women and elderly people), criminality and insecurity, problems of loitering youths, decay, and pollution of streets.

The neighbourhood approach increased cooperation between the city district, professional welfare providers, housing corporations, and residents, which led to many initiatives. One initiative of city district civil servants was to form a "social cohesion think-tank", wherein a group of active and successful residents was invited to join. This initiative for a think tank with citizens was meant to break the traditional division of roles between government, welfare organisations, and citizens. Active policies stimulated citizens to contribute to both policy making and implementation (Sterk 2011). Boluijt (2011, 1) notes that "*this ambition [for cooperation] is not new ... but the manner in which different organisations cooperate with residents and the city district is*".

The neighbourhood community was envisioned as a (constructed) collective of proactive and responsible citizens that join around the common purpose of improving the neighbourhood and community building. They would connect through networks of residents, (social) entrepreneurs, representatives of welfare organisations, and citizen initiatives. In other words, the neighbourhood community exemplifies making use of the strength of residents and citizens and building civil society (Sterk 2011). Between 2008 and 2010, three neighbourhood communities developed in the area: Timor square community, Karrewiel community, and Makassar square community. Their core business is to tackle urgent problems and really felt needs in the neighbourhood, making use of all resources available such as practical (residential) experts, professionals, entrepreneurs, creative citizens, peer groups, and formal entities such as local government, associations, educational centres, and housing corporations.

The Makassar neighbourhood community thus emerged in an enabling environment with government policies promoting active citizenship, self-reliance, participation, and integration of migrants. Hence, there was some funding available for facilitating citizen initiatives and participation and residents formed residents' panels to oversee the spending of those funds. The city district even created a specific civil servant function to stimulate and facilitate citizen participation. This so-called participation mediator has played a crucial role in the growth and success of the neighbourhood communities:

> "Participation mediator is a frontline position. I am part of the daily lives of the people in the neighbourhood and I am also part of the city district's system. This position resulted from the Neighbourhood Approach." (Interview with a participation mediator, June 2012)

The Makassar community is a vibrant and diverse group of loosely organised citizens that is still in development. It has formed a core group of 18 volunteering residents who coordinate activities and external relationships with city district, housing corporations, the central city council, and similar. This core group acts as a bridge between the formal organisations with their policies and regulations and the needs and aspirations of residents. However, there is a felt need to formalise and professionalise in order to deal with growth, diversity, and internal tensions. Already, the community has seen the need to create a legal foundation in order to be a serious interlocutor and partner of the city district. This led to intense internal discussions about representation, independence, transparency, and legitimacy.

Moreover, concerns about growth and professionalisation led to questions regarding (future) roles, identity, relations, and funding. So far, the community relies on volunteers, local entrepreneurs, and housing corporations. And more recently the community has been exploring relations with the (semi-)corporate sector like Triodos Bank (a social bank), Rabobank (a cooperative), and national supermarket chains. Some core members argue that more funds, including subsidies, are needed to make a real difference. Until now, local government provided small funds for a few concrete citizen initiatives.

The Makassar community has arrived at a crossroad where they have been recognised by (local) government as an interesting player in terms of successful citizen participation. There may be opportunities to apply for government subsidies and further connect to government policies. Likely, this will lead to a call for further professionalisation in order to deal with the speed of policy processes and (requirements and procedures of) government subsidies. With this, the community (or, more accurately, its core group) runs the risk of disconnecting from its wider base of a diverse, vibrant, and informally organised broader community and a civil society with its own processes and dynamic. It thus risks becoming a traditional government-funded entity that has lost its constituency and thus legitimacy. The question is whether there is another road for the neighbourhood community to relate differently to (local) government and not lose its independence and its strong roots in civil society.

Dutch Council for Refugees

In essence, it is impossible to understand (the history of) VWN without taking into account its relationship with the government. This is not only clear from the fact that VWN was set up in 1979 (as a merger of different organisations) in direct response to a government request but also from the financial dependence of VWN on government subsidies. More generally, the development of the Council for Refugees can be regarded as a combination of three intertwined developments: the changing number of refugees entering the Netherlands, the changing political framing of refugees, and changes in government policy (regarding refugees as well as towards civil society organisations in general – see the five phases discussed above).

With regard to refugees entering the Netherlands, two issues stand out. The first relates to the initial large increase in the number of refugees and subsequent reduction in influx due to, among other things, successful implementation of Dutch government policies. In 1994, about 50,000 applications for asylum where filed; later this stabilised at some 20,000 per year until 2001. After 2001, the number dropped to between 10,000 and 15,000 per year. The erstwhile increase in number of refugees created many opportunities for a growth in service provision, resulting in more paid staff. After 1979, for instance, the number of refugees increased as 6,000 Vietnamese boat refugees sought shelter in the Netherlands. In order to provide sufficient support VWN grew from an organisation with approximately 35 employees to a large CSO employing more than 400 employees. The decrease of refugees in later years led to less income for VWN.

Second, in general the group of refugees has become more complex. VWN moved from rather straightforward Vietnamese boat refugees to more complex refugees from Africa and the Balkans. One extreme cause for complexity are the AMVs (minor/underage refugees). This peaked in 2000 (6.705). For several years, AMVs remained a large percentage of the refugee group (considerably more than 15 per cent) but their number dropped after 2003 (and the end of some African wars) to reach a mere 3 per cent in 2006 (Dutch Council for Refugees 2011).

In understanding the importance of the government in the history of and the challenges faced by VWN the development of the political framing of refugee issues is of importance. Next to the objective changes in complexity and numbers, political framing of refugees also changed. Where once it used to be something like noblesse oblige and an act of international solidarity, nowadays (although numbers are really down) refugees are seen as part of the larger immigration issue. In this politicisation of refugees (and thus the increasing problematising of the issue) also the objective difference between (political) refugees and economic immigrants has evaporated.

Aside from these direct refugee-related developments, the development of the general civil society-government relations as indicated in the five phases described above is also crucial here. During the 1970s many CSO-like churches and activist organisations were involved in supporting and taking care of refugees. At the end of the 1970s, the then Minister of Culture, Recreation, and Social Work announced the government's intention to subsidise the reception of refugees (although restricted to those who had already obtained a residence status), but only if the various organisations would merge into one overarching organisation. Only this entity would have direct contact with government and receive government subsidy. As a result, VWN was founded in 1979, originating from a number of different organisations with diverging backgrounds. It provides a clear example of government taking over the governance function of organisations and serving as legitimiser for the merger.

The founding of VWN was a starting point for a long, frequent, and close interaction with the Dutch government providing a clear illustration of the interlacement between civil society and the state. By subsidising the work of the Council, the government had a strong hand in shaping activities and influencing decision-making of the organisation. Two issues are highlighted here to provide an elaborate picture of the relationship between VWN and the government. The first refers to the influence of policy on changes in the organisation and responsibilities of VWN, and the second concerns the challenges that a combination of roles creates for VWN – particularly also in its dealing with the government.

Two examples suffice to illustrate the importance of policy changes for VWN. In the early 1980s, and in contrast to government's idea that its subsidy was only meant to support refugees with an official residence status, VWN also used the subsidy to support asylum seekers. In fact, it expected funding to flow back into the organisation after this group obtained a status. In reality, however, many statuses were declined, leaving the organisation with large debts. Moreover, arranging accommodation for the rapidly increasing number of refugees became a heavy burden. In effect, the organisation decided to hand over the responsibility of arranging accommodation back to the government, cutting back the number of employees to less than 20 in 1981. It meant that in a very short period the size of the organisation increased rapidly, followed by a drastic decline again.

But not only subsidy preconditions led to changes in VWN. The introduction of the ROA Act (*Regeling Opvang Asielzoekers*) in 1987 meant that the accommodation of refugees was spread over the entire Netherlands, making local governments responsible for accommodation and integration. As local governments lacked knowledge on how to facilitate groups of refugees, VWN offered its support and expertise and started many local chapters providing back-up services while working in close cooperation with local governments. Eventually, their numbers grew from approximately 40 to more than 400 at the end of the 1980s. In addition, in the early 1990s procedures (for obtaining a residence status) were standardised and VWN agreed with the Ministry of Justice to support and counsel refugees in local reception centres. These new activities demanded specific knowledge on legislation and jurisprudence, which led to a process of professionalisation at VWN. Together with this increase in paid staff and a growing demand for high quality services and counselling, a focus on efficiency and effectiveness became important. Or as a staff member of VWN (Interview, March 2012) puts it: "*At the*

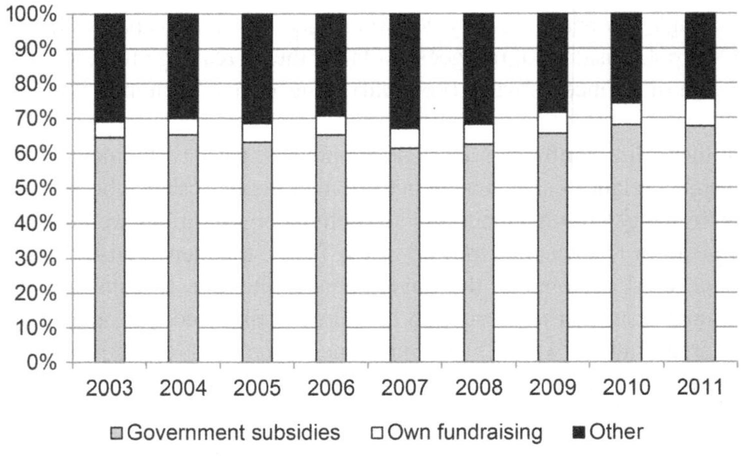

Figure 1: Origin of VWN's budget (in %, 2003–2011)

moment we are dealing with a continuing professionalization ... local municipalities have increasing demands: we need to be result-oriented".

The above demonstrates that VWN partly acts as the implementing agency of Dutch government policy and as such is also receptive to changes in that policy. This is also clear when looking at the dependence of the organisation on government funding. Figure 1 clearly shows that VWN receives more than half of its annual budget from government sources, principally for its contracting out-activities in counselling and integration of refugees in Dutch society. VWN also uses part of its other sources (e.g. own fundraising and, particularly, participation in the postcode lottery) for these type of activities, but uses these sources mainly to fund its lobbying activities towards the Dutch government. Although the organisation in financial terms thus strictly separates its counselling activities from its lobby activities, the combination leads to challenges that can be grouped under the heading of "not biting the hand that feeds you".

The case of VWN serves as an example of three of the five phases in the development of Dutch civil society. The start of the organisation marks phase 3 with the government forcing erstwhile small, more or less pillarised, organisations into one central big organisation. VWN used the systems and traditions created by that pillarisation to finance its organisation. Government, both at national and local level, used to have a tradition of supporting opposing civil society organisations with paid community organisers aimed at mobilising citizens to support the issue of refugees and the broader picture of international solidarity. This work is a combination of advocacy and practical help for refugees. Over the years, the practical help for refugees became a dominant part of the subsidised activities when VWN became officially involved in the central housing systems for people waiting for asylum. This created some extra tension between the advocacy and service parts of the organisation, with advocacy and awareness-building funded by private money (postcode lottery) and volunteering and service delivery funded by government subsidy and private volunteer support.

In phase 4, the Dutch government plays three different roles in its relation with the Council for Refugees: financing, governing, and contracting. First, it is a very efficient fundraiser, understanding that collecting taxes is easier and cheaper than civil society fundraising. Second, the government acts as a dominant stakeholder and board member involved in the governance of the organisation, with direct influence on VWN's activities. The most common way here is using the strings attached to government subsidies, but at the same time tolerating a certain degree of organisational freedom. In that respect, legal force as an instrument is used very infrequently (in the Netherlands). Third, government contracts out, leaving very little independence for the organisation. To a certain extent it has turned VWN into an implementing agency of the government.

The development in phase 4, a growing dependence and decoupling from civil society of the Dutch Council for Refugees, is enlarged by the increase in the number of refugees, changes in the type of countries of origin, and political framing of refugees. VWN becomes the plaything of these developments it cannot influence. As a consequence, for the Council for Refugees the general need to reconnect with their constituencies is enlarged. This makes VWN an interesting case to look at because next to the need to reconnect for financial and legitimacy reasons, the organisation needs to redefine its mission and its operational activities, both at national as at regional level.

Conclusion and discussion

The previous discussion portrays the Netherlands as a country with a long history of state-civil society relations. This relation has developed from an independent situation, where government funded civil society organisations, into a situation in which the difference between government and civil society in many fields has disappeared. Contemporary developments aim at recreating a more or less independent and voluntary civil society. The two cases provide an example of these new aimed developments (Makassar) combined with an organisational level story of what can happen when a civil society organisation develops extreme close, but less stable than expected, ties to the government (Refugee Council). While the relationship with government enabled a rapid growth of the Refugee Council, it also issues a warning for government taking control (phase 4), and retreating again (phase 5).

The Refugee Council represents the first crossroad where they find out that government subsidy leads to outsourcing of legitimacy and governance. They are at the crossroad of regaining their independence or becoming an organisation with a totally imposed strategy. Makassar represents the second crossroad where the organisation has to decide whether they want to enter the public funding system or not. Entering it may lead to growth and sustainability, but also to being pacified and controlled (co-opted in the system). If Makassar decides not to accept government funding, it is likely that they find out that they are unable to grow and could even become frustrated or hindered by (local) government. Being able to bend the rules, which seems to be the practice of Makassar, is allowed in this early stage of being small, unique, and below the radar. Once being asked to apply for subsidy, refusing the subsidy or not applying then becomes a dangerous signal. The latter suggests the often-expressed idea about Dutch civil society that it does not count unless it is subsidised by government. Although this might perhaps be difficult to substantiate, two other insights or lessons can be drawn from the Dutch case.

The first of these lessons concerns the government-financed national system of starting organisations, a characteristic of pillarisation. At the high time of pillarisation (mid twentieth century) the idea was that if in one pillar a civil society initiative was recognised by government and got subsidised, the other pillars would almost automatically be eligible for subsidy too and would thus be able to start the same kind of organisation. This happened in sports (e.g. football was played on Saturday by Protestants and on Sunday by Roman Catholics), recreation (e.g. scouting with at least five national associations), welfare, etc. This means that many organisations have a history of being rolled out from a national government funded office that implemented government favoured policies (supply-driven) instead of policies favoured by local chapters (demand-driven) – e.g. the failed roll out of "*feminist values*" in Scouting (Meijs 1997). The consequence of this all is that the central, national headquarters of multilevel organisations have limited income from local chapters and their policies have limited legitimacy in the eyes of these chapters. When, in phase 5, the government stopped funding these national headquarters, local chapters neither showed much compassion nor took over supplying this funding (see Meijs 2004).

Nowadays, and despite the fact that pillars have disappeared, the central, national approach of implementing welfare projects and organisations is still there, even within this WMO that was meant to create local freedom there are many tendencies to create national programmes. The Makassar case shows the influence of "*constructed ideas*" where the practice seems to be that professionals working with large welfare organisations in a community or national umbrella organisations assist in creating and setting up organisations in their own communities. Indeed, although this is much more like a grassroots organisation, in many cases the start-up is already based upon the idea of being or getting subsidised by government. Formalisation and professionalisation will be likely to lead to uniformity and loss of diversity and vibrancy that characterises the community. It is thus actually being counter-productive in terms of civil society strengthening.

The second lesson relates to the idea of subsidising civil society in order to pacify tensions in society. Here, the five phases described above show that during the times of pillarisation this pacification was very clear and successful. In more modern times, however, this is less clear as the elites of the pillars are disconnected with the disappeared "bottoms". The Netherlands are moving from a vertical pillar country, bridging (to some extent) class differences within pillars, into a horizontal, class-divided country (RMO 2010, 2011). Perhaps the pacification and control perspective should nowadays be understood in terms of (local and national) government financing the major part of paid staff, exercising control over the concerned organisation.

In service delivery organisations this type of control is combined with the New Public Management approach and has led to a downplaying of private norms and values in service delivery and the upgrading of neutral professional approaches (also see O'Neill 2002; Elbers 2012). In campaigning organisations this control is combined with an idea of consensus. It led to civil society organisations raising their voice in a very polite way (e.g. Dutch labour unions seldom go on strike while Dutch employer unions accept the right to strike) and framing their voice in more or less acceptable ways. A useful example is "Stop the Child Murder" (*Stop de Kindermoord*), an organisation asking for separate bike lanes and safe places to play. After getting subsidised, it downplayed its actions and even changed its rather aggressive name because paid staff did not want to have this rather "*controversial*" name on their business cards (Meijs 1997). Another point is that by subsidising only a central office of a diverse set of organisations, these organisations need to form one unified opinion amongst themselves before they can engage with government, as was already the case with the forced merger that created the Dutch Council for Refugees.

Note

1. http://english.minvws.nl/en/themes/social-support-act/ Accessed April 18, 2011. Municipality of Voorschoten, Social support act. http://english.voorschoten.nl/social_support_act Accessed June 25, 2013.

References

Bax, E. 1988. *Modernisation and Cleavage in Dutch society: A Study of Long Term Economic and Social Change*. Groningen: Universiteitsdrukkerij.

Boluijt, B. 2011. *Makassarplein community: tussen twee werelden*. Tilburg: Tilburgse School voor Politiek en Bestuur.

Burger, A., and P. Dekker. 1998. "De grootste nonprofit sector ter wereld." *ESB* 83 (4181): 944–6.

CIVICUS/De Nieuwe Dialoog. 2006. *CIVICUS Civil Society Index Report for the Netherlands*. Amsterdam: De Nieuwe Dialoog/CIVICUS.

Dekker, P. 1998. "Nonprofit Sector, Civil Society and Volunteering: Some Evidence and Questions from the Netherlands and the Rest of Western Europe." *Third Sector Review*, Vol. 4(2), 125–43.

Doorn van, J. 1984. "Inleiding: Anatomie van de Verzorgingsstaat." In *De interventiestaat*, edited by J. W. de Beus and J. A. A. Doorn van, 9–24. Meppel/Amsterdam: Boom.

Dutch Council for Refugees. 2011. *Vluchtelingen in getallen 2011*. Amsterdam: Dutch Council for Refugees.

Elbers, W. 2012. "The Partnership Paradox – Principles and Practice in North-South NGO Relations." PhD diss., CDIN, Nijmegen.

Hupe, P. 1993. "Beyond Pillarisation: The (Post-)Welfare State in the Netherlands." *European Journal of Political Research* 20 (23): 359–86.

Hupe, P., and L. Meijs. 2000. *Hybrid Governance, The impact of the nonprofit sector in the Netherlands*. Rotterdam: Erasmus University Rotterdam.

Kendall, J., and H. K. Anheier. 2001. "Conclusion: The Third Sector at the Crossroad? Social, Political and Economical Dynamics." In *Third Sector Policy at the Crossroads: An International Nonprofit Analysis*, edited by H. K. Anheier and J. Kendall, 228–250. London: Routledge.

Meijs, L. 1997. "The Regional Level in National Volunteer Organisations." In *Proceedings ARNOVA, Association for Research on Nonprofit Organizations and Voluntary Action*, edited by A. K. Plotinsky, M. Jackson and S. Dally, 124–5. Indianapolis: Indiana University Center on Philanthropy.

Meijs, L. 2004. "Changing the Welfare Mix: Going from a Corporatist to a Liberal Nonprofit Regime. Accessed April 26, 2013. http://www.jhu.edu/~istr/conferences/toronto/workingpapers.

Meijs, L. 2007. "When the State Sneezes, the Nonprofit Sector Catches a Cold: Reinventing Nonprofit in the Netherlands." In *Differing Images: The Irish Nonprofit sector and Comparative Perspectives*, edited by G. Donnelly-Cox and C. Breathnach, 39–50. Dublin: The Liffey Press.

O'Neill, O. 2002. *A Question of Trust – the BBC Reith Lectures 2002*. Cambridge: Cambridge University Press.

Pennings, P. J. M. 1991. *Verzuiling en ontzuiling: De lokale verschillen*. Kampen: Kok.

Pfeffer, J., and G. Salancik. 1978. *The External Control of Organisations: A Resource Dependence Perspective*. New York: Harper and Row.

Powell, W., and R. Friedkin. 1987. "Organisational Change in Nonprofit Organisations." In *The Nonprofit Sector: A Research Handbook*, edited by W. Powell, 180–94. New Haven: Yale University Press.

RMO. 2010. *Naar een open samenleving? Recente ontwikkelingen in sociale stijging en daling in Nederland*. The Hague: Raad voor Maatschappelijke Ontwikkeling.

RMO. 2011. *Nieuwe ronde, nieuwe kansen. Sociale stijging en daling in perspectief en de resultaten*. The Hague: Raad voor Maatschappelijke Ontwikkeling.

Salamon, L., and W. Sokolowski. 2004. *Global Civil Society: Dimensions of the Nonprofit Sector (volume 2)*. Bloomfield, CT: Kumarian Press.

Schulpen, L. 2006. "The NGDO Battle – Changes in the Subsidy Structure." In *Netherlands Yearbook on International Cooperation*, edited by P. Hoebink, 33–44. Assen: Van Gorcum.

Schulpen, L. 2009. "From 'Licking Wounds' To 'Private Initiatives' – Dutch Private Development Cooperation in 2007." In *Netherlands Yearbook on International Cooperation*, edited by P. Hoebink, 209–220. Assen: Van Gorcum.

Sterk, M. 2011. "Buurtinitiatief in nieuw perspectief. Een kwalitatief onderzoek naar de Karrewiel Community in de Indische Buurt in Amsterdam-Oost." Amsterdam: Master's thesis, Vrije Universiteit.

Stubbe, W. 2006. "Global and Local Behaviour of Nonprofit Organisations." PhD research proposal, RSM Erasmus University, Rotterdam.

The authors

Rik Habraken (corresponding author) works as junior researcher/lecturer in development studies (Centre for International Development Issues – CIDIN) at the Radboud University Nijmegen. He is responsible for the Dutch NGO-database and for the cooperation with university partners in Indonesia and Ghana. His topics of interest are civil society, development cooperation, and the aid architecture.

Lucas Meijs is professor Strategic Philanthropy with the Erasmus Centre for Strategic Philanthropy and the department Business-Society Management of the Rotterdam School of Management, Erasmus University. He specialises in philanthropy, volunteering, civil society, and business community involvement. He is a member of the Dutch advice council for societal development (Raad voor Maatschappelijke Ontwikkeling).

Lau Schulpen is senior researcher/lecturer in development studies (Centre for International Development Issues – CIDIN) at the Radboud University Nijmegen. He specialises in civil society, development cooperation, and the role of new actors in development. From 2006, he coordinates the IS-academy on civil society – a partnership between Radboud University and the Dutch Ministry of Foreign Affairs.

Cristien Temmink worked at PSO Capacity Building in Developing Countries, an association of Dutch development NGOs, until the end of 2012. She facilitated international learning processes and action-research on civil society, (endogenous) capacity development and learning-oriented planning, monitoring, and evaluation (PM&E). Her topics of interest are civil society development and relationships in complex processes of social change.

Treading new ground: a changing moment for citizen action in Greece

Maro Pantazidou

This paper looks at moments of citizen organisation to evoke an understanding of the new civic practices they bring to the fore. It argues that the Greek movement is shaped by an unprecedented number of citizens moving away from traditional, representative, recognised forms of citizen organisation to citizen-led, anti-hierarchical, horizontal networks that resist the consequences of the economic crisis and create alternatives to the current democratic and economic model. This movement has been nurtured by both small, diverse citizen initiatives, and mass protest and participation. The paper calls for a better understanding of the interplay between everyday resistance and significant political events.

Hors des sentiers battus : un moment transformateur pour l'action citoyenne en Grèce
Cet article traite de certains moments d'organisation citoyenne pour évoquer une manière de comprendre les nouvelles pratiques civiques qu'ils font ressortir. Il soutient que le mouvement grec est façonné par un nombre sans précédent de citoyens qui s'éloignent de formes traditionnelles, représentatives et reconnues d'organisation citoyenne vers des réseaux horizontaux, antihiérarchiques et dirigés par les citoyens, qui résistent aux conséquences de la crise économique et créent des alternatives au modèle économique et démocratique actuel. Ce mouvement s'est nourri de petites initiatives citoyennes diverses, mais aussi de la participation en masse. Cet article demande l'amélioration de la compréhension de l'interaction entre la résistance au quotidien et les événements politiques importants.

Abriendo caminos nuevos: una cambiante coyuntura para la acción ciudadana en Grecia
El presente artículo examina las distintas coyunturas de organización ciudadana con el objetivo de comprender las nuevas prácticas cívicas que dichas organizaciones ponen en juego. Sostiene que, en Grecia, el movimiento se caracteriza por el creciente número de ciudadanos que abandonan las formas de organización ciudadana tradicionales, representativas y reconocidas, ingresando a redes ciudadanas, no jerárquicas y horizontales, que resisten las consecuencias de la crisis económica al mismo tiempo que crean alternativas al actual modelo económico y democrático. El movimiento se ha visto beneficiado tanto por las pequeñas y diversas iniciativas ciudadanas como por las participaciones masivas. El artículo aboga por comprender mejor la interacción que tiene lugar entre la resistencia cotidiana y los eventos políticos significativos.

Trilhando um novo caminho: momento de mudança para a ação de cidadãos na Grécia
Este artigo examina os momentos da organização de cidadãos de promover um entendimento das novas práticas civis que eles suscitam. Ele argumenta que o movimento grego é influenciado por

um número sem precedentes de cidadãos que estão se distanciando de formas tradicionais, representativas e reconhecidas de organização de cidadãos e indo de encontro a redes lideradas por cidadãos, anti-hierárquicas e horizontais que resistem frente às consequências da crise econômica e criam alternativas ao modelo atual democrático e econômico. Este movimento tem sido gerado tanto por pequenas e diversificadas iniciativas de cidadãos quanto pela participação da massa. O artigo clama por um melhor entendimento da interação entre a resistência do dia a dia e eventos políticos significativos.

Introduction

On 23 May 2011, I was sitting in front of the Spanish Embassy in Athens surrounded by around 30 or 40 people. The Indignados movement was filling the squares of Spain and every day, Spanish residents of Athens staged "real democracy" demonstrations in front of their embassy. A few Greeks had joined them. Looking at the people around me I complained to a Greek friend next to me: *"Why don't we see a mass mobilisation of people here? There has been a year of austerity packages that lead nowhere, what are people waiting for?"* His response was sharp:

> "You are looking for a mass movement and not paying attention to the things happening below the radar. People's lives are changing so fast, we are in shock, urged to reconsider our democracy while wondering how to make ends meet. But people are organising, new activity will come. The movement develops slowly but it does."

Two days later a spontaneous gathering of 200,000 people turned into the Greek movement at Syntagma Square, which was then occupied for more than two months.

My friend was right: before and after the Syntagma Square Movement (SSM), citizen agency was present, channelled in a number of smaller-scale mobilisations and initiatives in which a new culture of social change was and is being cultivated.

In this article I take a closer look at these moments of citizen organisation to evoke an understanding of the new civic practices they bring to the fore. I argue that the Greek movement is shaped by an actual movement of an unprecedented number of citizens away from traditional, representative, recognised forms of citizen organisation to citizen-led, anti-hierarchical, horizontal networks that both resist the consequences of the crisis and create alternatives to the current democratic and economic model. This movement has been nurtured both inside small, diverse citizen initiatives and during moments of mass participation, such as the SSM and the 2008 Greek riots. Acknowledging this, the article calls for a better understanding of the interplay between everyday resistance and significant political events.

I suggest that at the heart of this political creativity are solidarity, unruly action, direct democracy, and autonomy, which drive and shape the claims, discourses, practices, and processes of emergent citizen action. These driving forces open possibilities for new relationships among citizens and between citizens and institutions, setting the stage for a new politics in the current European situation of increasing poverty, unemployment, and corrosion of the social state and democratic practices.

This article is based on desk research of stories, actions, and analysis from citizens in Greece, and on interviews and group discussions with individuals active in various citizen initiatives. It is also informed by theoretical insights on citizen action, politics, and movements. It starts by providing a brief introduction to the Greek context and to the methodological pathways of the research. The first section discusses the different faces of citizen action in the age of the Greek crisis by examining both the role of mass protest events and the dynamics of emergent smaller-scale citizen initiatives. Instead of offering conclusions –unsuitable for a rapidly changing context – the second section reflects on the contributions of the Greek case to our understanding of new civic practices and social change.

Entering the scene

> "There is an urgent need to deconstruct the moral lessons that obscure the actual process at work in society." (Alain Badiou 2005)[1]

Greece has been at the epicentre of the European debt crisis, and the site of the fastest structural adjustment programme witnessed in an EU country. The crisis has incurred dramatic transformations at the socio-political and economic levels and has brought the whole society to a crossroads.[2] Statistics cannot capture neither people's realities nor the psychological pressure exerted by media and political parties, who have repeatedly asserted the so-called absolute necessity of emergency measures.

Over the last two years, Greek governments have agreed two loan agreements with the "troika" formed by the IMF, the European Union, and the European Central Bank. These loans are attached to a set of conditionalities, which include the implementation of heavy taxation, rapid deregulation in the labour market and health sector, privatisation of state property and natural resources, and cuts to state services.

Since Greece started implementing the conditionalities in 2010, unemployment has risen from 10.7 per cent in January 2010 to 22.5 per cent in April 2012 (1,147,372 unemployed in a country of 11 million), while figures for youth unemployment reached a historical high of 52.8 per cent in April 2012. Labour rights have been severely weakened through a new set of laws that abolished collective bargaining and withdrew restrictions on companies' rights for redundancies. The national minimum salary has gradually[3] been reduced, pulling both public and private sector salaries into a downward spiral. All direct and indirect taxes have also gradually increased. Five per cent of the Greek population (more than 500,000 people) have seen their incomes fall below the poverty line, increasing to 25 per cent the portion of the population living in poverty (Matsaganis and Leventi 2011).

These measures have primarily targeted the working population and lower and middle classes, reducing their living standards while at the same time depriving them of basic mechanisms for social protection. The welfare state has been blatantly dismantled, with severe cuts to pensions, unemployment, and disability benefits, and the introduction of fees at public hospitals. These policies resulted in large categories of the population struggling to cover their basic needs. It also brought about a rapid deterioration in the provision of health care and education, the collapse of thousands of businesses, and an increase in mental illness, with Greece experiencing the fastest-growing suicide rate in the world (Economou et al. 2011).

Whose democracy?

The fierce attacks on social and economic rights have occurred alongside a corrosion of democracy and social cohesion. As previously seen in developing countries, the structural adjustment programme was accompanied by the derailment of democratic processes: an unelected technocrat government was appointed in November 2011 and crucial bills that fundamentally alter citizens' rights were passed using fast-track processes – often with little or no deliberation in parliament or public consultation.

Police repression was increased to control public outrage. Freedom of expression was also affected. Greece fell sharply in the 2010 World Press Freedom index, to bottom in Europe, in 70th place (it ranked 35th in 2009), The country has also seen the fascist Golden Dawn party dramatically increase its power: following the last election it has 18 Members of Parliament, while its militia attacks and harasses immigrants on the streets (Human Rights Watch 2012).

Traditional civil society has attempted to respond to the threats and consequences of the crisis. Trade unions have organised more than 17 national strikes in two years and a small number of NGOs have come up with renewed service delivery projects or anti-racist campaigns to address the rise of fascism. These actions, though important, haven not yielded political results or attracted the genuine participation of a majority of citizens in constructing solutions and alternatives. Interesting enough, citizens are moving beyond the traditional mechanisms of participation to construct their own alternatives for addressing each facet of the crisis. These alternatives are further explored below.

Methodological pathways

> "The movement is not (yet) an 'object' of study. It is a lived experience and to speak about, understand or judge the 'phenomenon' you need to have 'lived' it." (*Drasi Newspaper*, April 8, 2012)

The quote above highlights the theoretical, methodological, and ethical challenges of objectifying and studying social phenomena as they are evolving. In a period of heightened interest in studying and understanding the diverse expressions of citizen agency around the world, I felt it was important to discuss the notions and processes I navigated and the challenges I had to keep in mind while researching this article.

Looking for civil society inside the movement . . . or is it the other way round?

As this research was instigated by the Civil Society at Crossroads initiative (www.pria.org/civil-society-at-crossroads-), an obvious question arose: How does the concept of civil society fit with emergent, not yet labelled forms of citizen action? I chose to work with the understandings and notions of civil society that were relevant to the active participants of the Greek movement.

As in most of the movements that swept the world in 2011, traditional civil society actors didn't play a crucial role in developing new civic practices in Greece. Instead, citizens

Table 1: Understandings of civil society, developed in discussion with young activists

A	vs.	B
Social projects and programmes		Social change processes
Tasks entrusted/assigned to leaders and experts		Self-organisation and direct democracy
Recognised and socially acceptable forms of participation and claim making		Unruly, disruptive or illegal direct action
Negotiating specific problems and interests – depoliticising		Addressing multi-faceted crisis and putting forward holistic claims – politicising

turned to informal, spontaneous, self-governed forms of organisation. This can be explained, when we consider the divergence between the project-based culture and the norms of hierarchy and expert knowledge that characterize many formal organisations and the radically political, non-representative, and inclusive ethics of the new citizen initiatives.[4]

Taking into account the limitations to the concept of civil society and the dynamics of grass-roots action under construction in Greece, I consciously looked for those spaces and events where citizen action is channelled and nurtured beyond formal structures. My entry point was the question: what forms of participation did people look for when they felt the need to share their concerns and defend their rights and needs? When one looks for answers to this question, a world of collectivities,[5] citizen initiatives, and acts of disobedience and solidarity is revealed – which some research participants argued is the "real civil society". Table 1 was developed to help clarify these two competing understandings of civil society during a group discussion with young activists.

As one group discussion participant observed, the polymorphous galaxy of citizen initiatives in Greece constitutes a movement because it signifies a transition – a literal movement – from the "A" understanding of citizen organisation and action towards the "B": people are rejecting or reshaping forms of civic association that are based on the A and are situating themselves as bodies and minds in the public space with the intentions to engage with the B. This article aims to understand the dynamics of this movement and the practices, claims, and discourses it brings to the fore.

Methods and process

Attempting to understand these dynamics in a rapidly changing context is a challenge. For this reason and to be consistent with the "emerging phenomelonogical" approach of the Crossroads initiative, I chose a snowball approach using people's experiences as an entry point. I focused on conversations with people who have been directly involved with citizen action; they provided leads to different initiatives and analyses of the dynamics of the movement that they deemed most important. Those leads and analyses led me to consider initiatives and actions that I was unaware of or had not paid significant attention to, developing conversations and information that helped create a rich picture of the movement.

In order to analyse this rich picture, my methodology was iterative, and based on informal conversations with a number of actors.[6] Specifically:

(a) interviews with members of six different initiatives that reflect different faces of the emergent citizen struggles: a neighbourhood assembly, a workers' association, a "disobedience" movement, a resistance network, one solidarity, and one alternative-economy initiative. The interviews took the form of informal conversations to allow for bottom-up and collaborative sense-making of events. The questions were loosely structured around the activities of each initiative, their goals and motivations, difficulties and limitations, as well as their broader take on the current movement in Greece.

(b) two group sessions/small workshops, one with representatives from eight prominent NGOs and one with young activists who were part of the SSM and are currently involved with different citizen initiatives. Here, my goal was to create a dialogue in which everyone (including me) was asked to share their observations on the changing forms of citizen action. Lessons were drawn through collective reflection and critical analysis.

The findings of the discussions with the protagonists of the action were complimented by collection and analysis of other case studies, think pieces, posters, slogans, and social media content. These came from desk research, the snowball effect, and my own archives from having followed the crisis over the past three years.

By collecting stories, facilitating analysis with diverse groups of people, and using content analysis, the article aims not for triangulation – as there is not one state of citizen action to be discovered – but for "*crystallisation*" (Richardson 1994), trying to capture the dynamics from multiple perspectives. Because "*in a participative worldview, we always partake in what we describe*" (Reason 1994, 324), this article is also largely influenced by my multiple roles and identities as a researcher, facilitator, activist, Greek citizen, and NGO practitioner. I stand in solidarity with the struggles of the people I engaged with during the research and in some cases, I perceive myself as standing to gain from these struggles. I chose not to hide this crucial subjectivity and made sure to create conversations with research participants in which both sides exposed their views and thoughts rather than one in which I asked objective questions about some distant situation. This helped create trust among participants and encouraged open dialogue. To honour this trust and for reasons of privacy and/or safety, participants' names are initialised, as are some of the names of the collectivities and initiatives.

Treading new ground

What civic practices have developed in Greece during the years of crisis? What is the significance of political events and what is the content and role of the various ongoing citizen initiatives? What is the interplay between events and everyday action? This section provides answers to these questions and paints a rich, yet in no way exhaustive, picture of the new grounds citizens are treading.

This first part discusses two moments of mass citizen participation: the 2008 riots and the Syntagma Square Movement of summer 2011. Although the latter is widely recognised as the culmination of Greek resistance to the policies of the troika and the Greek government, both events played a crucial role in embodying and consolidating emergent elements of political culture. At the same time, they laid the ground for new civic practices.

The second part presents and explores the modalities, goals, and realities of new civic practices through the faces of citizen action that underpin them: solidarity, direct democracy, unruly action, and autonomy.

Background to the events

The post-WWII history of the Greek state is dominated by the power of political parties. Such has been their influence in all domains of social life that academics have talked about "*partitocracy*" and "*colonisation of the associational sphere*" (Mouzelis and Pagoulatos 2002). Trade unions, student unions, social services, and even local sports and cultural associations were, and to a large extent still are, affiliated with a political party. Some of the first organised movements for women's rights and environmental issues in the 1970s and 1980s also occurred under the umbrella of the main political parties. Clientelistic practices were gradually and deeply ingrained into all sectors of public life, and, in combination with Greek society's strong traditional kinship bonds, in many ways replaced the role of a functioning social state and a strong civil society.

The grip of parties loosened somewhat beginning in the late 1980s, to be replaced by the advancing power of the markets and the influence of a minority elite which controls the media and business sectors (Mouzelis and Pagoulatos 2002). This period saw also the formation of both national and local NGOs working on human rights, environmental, and cultural issues, and in some cases providing services to the most vulnerable. The growth of NGOs, the promotion of EU-funded active citizenship programmes, and the 2004 Athens Olympics gave a boost to a culture of volunteerism, although, in general, participation in NGOs has not been widespread in Greece.

The eruption

In addition to "partitocracy" and later "elitocracy" and the low interest in NGO-type action, or indeed because of it, Greek politics have always been characterised by a strong presence of popular protest movements and revolts. These movements define themselves as antagonistic to the state and to neoliberal policies.

Over the last two decades, three mass student movements (1991, 1999, and 2007) occupied hundreds of schools and universities across the country to resist education reform laws that were seen to be threatening the values of free and public education. National strikes are also a common phenomenon in Greece. The country has a strong tradition of anarchist and extra-parliamentary leftist collectivities that have often staged protests, clashed with the police, and in many cases led the anti-racist/anti-fascist struggle in the country.

Examining the dynamics, claims, and organisational tactics of these movements is beyond the scope of this article. But it is important to point out that the impressive number of protests, strikes, clashes, and occupations that Greece has witnessed in the post-dictatorial period (since 1974) indicate the wide distance between those who govern and those being governed. The lack of genuine channels for citizen participation and the marginalisation of forms of political expression not connected to political parties has largely been compensated through clientelistic practices.

Much of this "manufactured consent" between society and the political establishment collapsed in the days following 6 December 2008. That night, a police officer murdered, in cold blood, a 16-year-old boy in the Exarchia district of Athens. The event sparked three weeks of unprecedented riots all over Athens and the main cities of Greece. Thousands of people who normally have no role or space in co-shaping politics – young unemployed and precarious workers, marginalised immigrants, school students – immersed themselves in spontaneous direct action. Rioters looted and threw stones, and wrote texts and messages on walls, printed posters and leaflets, and blocked streets, creating an event that brought diverse people together to reclaim the streets as never before and reject the way politics were being run as a whole:

> "Before December, we knew it already – no one was to be trusted, politics was corrupt, things were getting irreversibly worse all the time and there was nothing to do about it. But then we took to the streets, we found each other, and there was actually no need to read what other people wrote and do what other people had arranged to do and wait for others to think about what we want, no need to articulate demands and ask for marginal benefits so they could understand, no need to adopt argumentative strategies and representative ethics so as to reach a national consensus, no need to have a meaning within this frame, because we had no need of this frame, we create our own meanings ... no one was representative of any group, but everyone was represented, nothing of what we asked for could be articulated in the language of political demands but everything was said ..." (Kouki 2011, 171)

As Kouki demonstrates, the December riots opened up a new space for politics exactly because "*they escaped the language of those who have the power to define politics*" (Badiou 2005,). They created a moment of rupture with the way protest, anger, or demands had previously been expressed or were expected to be expressed. Messages against media propaganda, police impunity, and lack of democracy filled the walls of the city. The invisible people of the city made themselves the centre of politics, crying out "*we are here, we are everything!*" (Douzinas 2011, 160).

"December" was not a moment of explosion but of implosion: a system that could no longer sustain itself collapsed (Giovanopoulos 2010). As the riots faded in early 2009, less than a year before Greece started talks with the IMF, the protagonists of December thought that inevitably things would return to normality. But a new process of radicalisation had started in society. The ability of the governors to create consent had broken. December had triggered a new era of citizen agency: people started forming neighbourhood assemblies to discuss "what is happening in our lives", turning to social media for alternative information, and embracing the power they have to create political events on their own terms.

Change of repertoire

After December 2008, the Greek political system, having lost much of its legitimacy, responded with even more cynical political decisions. Any prior effort for social consensus was disregarded. Consecutive governments have imposed measures they themselves characterise as unfair and have used police repression and scare tactics to ensure compliance. Academics have talked about a "*meta-democracy*" (as noted in edromos.gr website, December 6, 2010; Douzinas 2011).

People started moving away from the old repertoires of protest that had become largely predictable and consequently easy for the powerful to manipulate (in framing public discourse) and to ignore (in making actual decisions). Citizens started rejecting old, representative, and professionalised tactics for organising and progressively engaged with a political culture of unmediated presence via direct democracy practices in assemblies and through social media. The drastic and shocking change of political regime after Greece went under troika supervision led to the Syntagma Square Movement (SSM), which lasted for two months between May and July 2011.[7]

The "movement of the squares" crystallised and further developed this new political culture:

> "It publicised the knowledge that was already being developed in streets, neighbourhoods and social centres – that there is a new way for us to organise from below, to discuss our problems, to find joint solutions."[8]

Crucially, on the first day of the movement, the Syntagma people's assembly demanded "direct democracy" instead of the "real democracy" claim put forward by the Spanish *Indignados*; thus relocating the focus of the struggle from demanding a fairer implementation of the current democratic system, to implementing another kind of democracy in which all people are equally responsible for decisions affecting their lives. Even more crucial was the fact that the protesters created a process that reflected the essence of their claim: the Syntagma Square camp was run by the daily people's popular assembly and a number of thematic groups (on economy, direct democracy, debt, etc.) and organisational teams (on food provision, safety, etc.) all directly accountable to the main assembly.

Participation in SSM gradually dwindled after three violent police attacks and when this vehicle for action had reached its full potential. The police evicted the remaining protestors on 29 July 2011. Some saw the end of the Syntagma occupation as a failure. But its practices, experiences, and modalities were trans-located and continued in other spaces and moments of solidarity, disobedience, and direct democracy that are debated below. Indeed, the SSM "*never left the square*" (as observed by Ch. Giovanopoulos during an interview).

Faces of citizen action

This section draws on conversations with members of solidarity and resistance networks, people's assemblies and alternative economy initiatives as indicators of the different faces of citizen organisation and action in Greece: solidarity, direct democracy, unruly action, and autonomy.

These faces of citizen action are not distinct from one another: they complement each other and co-exist within each of the different initiatives presented here. They are methodologies to use and goals to aspire to, and they form part of the practice and language of citizen initiatives. Solidarity is expressing the need to collectively address rising problems, as direct democracy is expressing the desire to infuse equality in and control over decision-making. Unruly action speaks to a lack of social justice while alternative economy practices attempt to create a fairer society through a new culture of production and consumption. They reflect the growing popularity of the idea that solutions to our problems need to be collective, and the methods we use need to reflect and prepare for the society to which we aspire.

Solidarity

> "If we don't model our vision in what we do, if we create in our collectivities the same relationships created outside (in the system), then we are just doing philanthropy." (C.M.)[9]

As local authorities, the Greek Church, and a small number of NGOs organise soup kitchens and clothes/medicine/food banks for the increasing number of people who cannot cover their basic needs, citizen groups are creating their own solidarity structures. These citizen initiatives are principally characterised by the relationships of equality that they aim to foster between those who participate in them, and by the rejection of the culture of social provision and charity.

One striking example is the "social kitchen" started by K.M. The kitchen cooks and serves free food for all on the street. Every day members of the kitchen carry all the necessary utensils to a different part of Athens. People and passers-by join in and cook and then eat together on the street. The lines between volunteers who come to support the kitchen and people who come in need for the food become blurred. In fact, a number of people who approached the kitchen in need for food became frequent cooks and organisers of it.

The social kitchen is suitably named "O allos anthropos – The Other Man": *anthropos* in Greek refers to the concept of the human being. More than providing food, it is about a process of resocialisation, where people from diverse backgrounds can discover a new way of being in the city and society as the streets become *our* home. It is also about placing oneself in the midst of the activity (cooking, sharing food, serving, and eating): "*we are asking for participation – not donations*" (K.M)[10]

The Other Man is just one of the many – simple in function but dense in meaning – collective kitchens that have sprung up across Athens. Those are joined by citizen networks that give or exchange clothes, social health centres run by doctors and citizens assisting them, and by citizen groups providing teaching assistance to kids, language training, and psychological support.

What makes these initiatives unique is that they are from citizens to citizens. As one of the coordinators of the Network of Solidarity and Resistance said: "*We are not in any way different than the people who come to take the clothes, we also struggle financially, are unemployed or might be tomorrow.*"[11] These initiatives constitute a citizen-led movement that is about dealing with the crisis through redefining the meaning of property and sharing, belonging and community; rather than donating or contributing to social support programmes run by government and agencies.

Direct democracy

> "How long should an assembly last when we need to work the next day? How will we cover the needs of the unemployed? How can we coordinate with other local assemblies and what are the limitations of lacking a general political strategy?"[12]

Questions such as the above have been at the core of many direct democracy initiatives growing in Athens and Greece. Citizens are resisting the deterioration of their lives in ways that pay due attention to both the process of organising and the form and content of the diverse actions that are needed to solve their problems.

Open Popular Assemblies (or neighbourhood assemblies) have been the main vehicle for rolling out this direct democracy experiment at the local level, both before and after the SSM. After and during SSM, neighbourhood assemblies multiplied. Currently, there are around 30 active assemblies in different neighbourhoods of Athens and another 30 in towns across Greece. As one member of an inner city neighbourhood assembly described:

> "After one week at Syntagma, people spontaneously started our local assembly. The neighbourhood is our space of responsibility for collective action that speaks to local but also to broader problems. Many of us were in Syntagma but we felt we could take more meaningful action at the local level; the Syntagma assembly voted for taking down capitalism and rewriting the constitution 10 times!"

So, in many cases, local assemblies were born of the need to take actionable decisions, directly relevant to people's lives, something that is often limited by the scope and size of national movements that situate themselves in the central political stage.

So what do assemblies actually focus on? Assemblies act as (a) a form of pressure on local authorities (resisting the commercial exploitation of public spaces, demanding support for the homeless, etc.); (b) a space for socialisation, communication, and the solving of immediate problems (supporting families left without electricity, creating green spaces, etc.); (c) a vehicle for local-level resistance against the current national economic and political system (occupying local tax offices to protest against taxes, awareness raising on national policies in local squares, etc.); (d) a space for the creation of new social and activist relationships as a demonstration of the future society members aspire to (adapted from Open Assembly of Citizens of Ag. Paraksevi 2011).

Different assemblies may state that they aim to fulfil one or many of these functions, or can embody one or all of them without necessarily stating it explicitly. The inner city neighbourhood assembly I spent time with was simultaneously taking or planning action to: ensure free access to the local hospital for all; plan a reach-out campaign to migrant communities; organise anti-fascist events; support indebted and low-income families, and citizens who are facing legal charges after participating in demonstrations.

Overall, the assemblies, like SSM, operate as self-governed communities according to the principles of direct democracy: they make their own rules and processes which are always open to review and change. All actions are agreed upon after long debates on their necessity, justification, and potential impact – there are no prescribed courses of action dictated by specific ideologies or methodologies for social change.

Unruly action[13]

> "These are actions legitimised in people's consciousness." (Research participant, member of neighbourhood assembly)

The years of the crisis have seen an increase in the number of people involved in forms of action that might appear disruptive, illegitimate, or illegal when opposed to the recognised norms of social mobilisation and the democratic (or one might say bureaucratic) routes for voicing public concern and pressure.

Such actions might aim to declare protest and exert pressure, like the occupation of the Syntagma Square or the relatively frequent and symbolic occupations of town halls and public buildings by their employees and trade unions. In other instances, acts of unruliness aim to reverse a perceived injustice in the present, albeit temporarily. As the crisis deepened, unruliness spread to other everyday interactions with the state: citizens are now occupying the tills at hospitals so that patients can be seen without cost, reconnecting the electricity in houses where provision was cut, preventing tax offices from printing foreclosure notes, and blocking ticket-issuing machines at public transport stations. These moments of disobedience are easily characterised as 'illegal or self-serving, but one needs to look beyond the discourse of the immoral citizen echoed by media and government alike. These actions appeal to another kind of moral economy – an underlying understanding of what is fair and just that permeates a society in a given context (Thompson 1971). When the state is understood to be acting in illegitimate ways, citizens step in to correct it, to show how they want to be governed (Hossain 2011). For example, the act of occupying the hospital tills to ensure free medical screenings is an attempt to show the state a fairer way the health system should work.

The case of the *charatsi* tax[14] is another example where "I don't pay" movements, assemblies, and resistance networks fiercely opposed the measure and widely disseminated information about why it is unconstitutional and how not to pay it, while at the same time establishing action squads that reconnect electricity to affected households. The issue received massive attention and the Constitutional Court deemed it unconstitutional to cut the electricity if someone fails to pay the tax, a partial victory for the movements.

This example showcases the hybrid function these initiatives perform. They set out to solve people's problems through direct action (as many citizens do not actually have enough to pay for tickets, tolls, etc.) and develop into awareness-raising and state-monitoring networks (citizens can reach members of the networks or visit their sites to learn their rights and see what each new government measure will mean for them in practice). In conclusion, direct and unruly action has for some become the only tool citizens have to demand accountability and correct injustices, but it is also increasingly used with programmatic intent by horizontal networks who want to oppose government policy and empower other citizens to engage in civil disobedience.

Autonomy

> "Considering the multi-faceted crises of the current moment we believe that there is an immediate need and a great opportunity for creating a network of collaborative and solidarity economic initiatives, through which we will re-define our everyday life, both in parallel to and in antagonism with the dominant system." (Founding Declaration, Nea Guinea collective: www.neaguinea.org)

Autonomy in Greek means self-governance. It refers to a society that decides on its own rules and processes, which are then constantly reviewed and monitored by the society itself. Currently, a number of citizen initiatives are focusing on creating self-organised and self-governed practices of production, consumption, and work in an effort to approach autonomous practices in the social and the economic sphere.

There are currently hundreds of citizen initiatives in Greece based on the ideas of solidarity or exchange economy, with much difference in terms of form, content, and focus. There are, for example, solidarity trade hubs that bring products directly from producers to consumers. These hubs operate along to fair-trade principles and at the same time encourage producers to form horizontal networks amongst themselves. There are also online "exchange economy" or "alternative currency" networks where people can exchange goods for services. Finally, there is a boom in city gardening, food sharing networks, and other practices that advocate self-management of food production and consumption.

There is a telling increase in the number of cases in which citizens occupy a green area, either in use or abandoned by public institutions or private companies, and set up a gardening and food production collective. Most of these initiatives state that their effort has a political character, and wish to creatively demonstrate an alternative of how natural resources can be collectively and democratically managed.

In another example of promoting autonomy in everyday life, the Nea Guinea collective in Athens organises workshops on the self-management of health, food, energy, and clothing. They are calling it a struggle to "*reclaim our lives*" through "*economically sustainable*" and "*socially just processes*" (Founding Declaration, Nea Guinea collective: www.neaguinea.org). The name Nea Guinea represents a utopia, a distant destination where solidarity and respect define human relations and humans' relations with the ecosystem.

Both Nea Guinea and other initiatives operate as workers' collectives. Workers' collectives constitute an interesting hybrid between citizen initiatives and labour formations. They are run by an assembly of workers, all paid the same hourly rate; profits made are used to support the growth of similar initiatives. The majority of these initiatives see themselves as part of the broader movement of direct democracy and resistance described in the sections above. They constitute the other side of unruliness, where injustices of the governed are not denounced through disobedience but are overcome through experimenting with the creation of different relationships and practices.

Reflections: a moment of change for citizen action in Greece (and Europe)?

As European peoples are crushed under harsh austerity, unemployment, and unrepresentative institutions, the following years might see rapid changes in the repertoires of citizen action in Europe as a whole. In this context, what is significant about the developments in the Greek terrain of citizen action? How do these developments contribute to our understanding of citizen action and social change? What follows are some preliminary reflections on these questions with a view to adding to the ongoing debate among scholars, activists, and practitioners concerning the changing face of citizen action.

Powerful practices

The Greek case shows that while democracy and the social state disintegrate, citizens take center stage and experiment with new civic practices. The new civic practices constitute both vehicles for resistance to attacks by the current socio-economic system and hubs of political creativity where alternatives to that system are created in the here and now. These are alternatives on how people can be organised, needs met, decisions taken, spaces used, production and consumption managed – or to put it more simply, they are new politics. These practices are shaking the conventional paradigms of how citizenship is lived, rights are claimed, and power contested. Their contribution is that they are introducing a new political culture and new mechanisms of contestation (Khanna 2012). Direct democracy, horizontality, and autonomy become the core of a vision of a different society – and this is a vision that seems to be modeled in the process of the struggle thus making the process more powerful; the process becomes the way to reach the goal.

In Greece, as in the Occupy movements in Europe and the US, democracy was not only expressed as a claim, but was reflected and enacted in the forms and the content of citizen action. Those forms – open assemblies and horizontal networks with loose leadership – rejected representation and chose unmediated presence, replaced the culture of assigning tasks with direct action, challenged expertise and opted for tacit and networked knowledge, overcame the logic of the majority and looked for consensus.

Although one can say that Greek political collectivities have long been familiar with direct democracy practices, over the last two years the latter have been rolled out and tested by an increasing number of citizens initiatives active in diverse areas of social life: labour, health, education, production. More and more, workers' and unemployed collectives, base unions, hospital committees, and school student boards are now operating through consensus-making assembly processes. What is particularly fascinating about this development is the commitment to a non-instrumental use of direct democracy which is not used just as a tool for achieving decisions but as an approach to how we can co-exist with our neighbours, colleagues, or fellow citizens within fairer social structures. At the same time, the practice of direct democracy poses its own challenges; its perceived openness and promise of equality might obscure the existence of power relations that divide any group along class or racial lines and so on. Additionally, a number of activists have pointed out that the consensus-seeking process and the horizontal nature of their networks might be slowing down or inhibiting the formation of a – according to them – necessary political front with a unified strategy.

In addition to the surge of direct democracy, the days of the crisis saw unruly and direct action embraced by a large number of so-called average citizens. Motivated by indignation that "*the absolute minimum conditions for securing economic and political justice are being violated*" (Tadros 2012), citizens are organising and promoting massive disobedience to taxes and laws and are "*rudely*" correcting an illegitimate state and market by blocking or taking over state functions (Hossain 2009). Again, in the case of unruly action, the reflection of the claim in the nature of the action, legitimises these practices in people's minds and reveals the illegitimacy of the mandates of the powerful. This is what is being achieved when, for example, rather than blocking a road to protest health privatization, citizens close down the hospital cashiers' office to allow patients to be examined by the doctors free of charge, thus reversing in the moment the injustice and demonstrating how a fairer system would look.

The growing body of experience with and exposure to these powerful practices can gradually serve as a pool of lessons and for both Greek and European movements.

The nature of social change

The new forms of citizen organisation and action might have been registered and debated as significant and innovative, but there has also been mounting criticism, and doubt about what this political creativity can actually achieve. What can citizen collectivities and networks such as the ones outlined here accomplish? To what extent can they affect change at the level of structural and systemic injustice in a country such as Greece?

I would argue that the formation, survival, and expansion of such initiatives has already achieved a great deal. They have moved political action beyond single-issue claims and put the underlying logics of the multi-faceted crisis we are facing on the agenda. They have created spaces where individual and collective change is taking place every day, where the values of solidarity-autonomy-democracy operate, and where tools that ensure a more liveable reality are created (Wainwright 2012). In these spaces the relationships between citizens themselves and between citizens and the institutions that govern our lives are transformed and reconsidered.

These developments invite us to re-examine what we understand as structural change and where we expect it to come from. Citizen movements in Greece, as elsewhere, are focusing on creating alternatives through small-scale networked action rather than through united national fronts that put forward lists with specific demands. This new form of citizen agency suggests that old models of exerting political influence might not be relevant any more – or at the very least might be a limited pathway to addressing complex global crises.

Additionally, the pathways along which new forms of citizen action emerged over the past years in Greece highlight the often-underplayed interrelation between mass centralised political events (such as the riots of 2008 and the SSM) and everyday decentralised citizen activity. Political events that might appear ineligible to recognised ways of doing political methods or that are understood as a one-off venting of anger are actually the arenas where new political cultures that will lead to new forms of action are created – as the Greek case showed, the lessons of riots and Syntagma spilled over into citizen initiatives.

Some discount these events as having no real democratic outcomes and defend the importance of everyday practices. Others overlook the struggles going on in the everyday just waiting for the next eruption. Movement dynamics in Greece suggests that there is a cycle, a continuum of citizen action with cycles and peaks, its ebbs and flows (Giovanopoulos and Mitropoulos 2011). Events are ruptures in the everyday that allow reality to be renegotiated. Through providing new practical, imaginary, or discursive resources (Khanna 2012), political events become the life-giving force or multipliers of new practices which in turn prepare the ground for new moments of mass participation in citizen action.

Will these faces of citizen action lead to sustainable changes in the political establishment and in civil society? Or will the new modalities be co-opted by the powerful or wither away once the crisis fades?

I feel that we are struggling to answer these very valid questions with our existing models and knowledge frameworks. Perhaps we are in need of new frameworks to capture how civic practices are diffused and adjusted, and to understand the interplay between events and practices, resistance, and alternatives. In this sense, the current moment stands as an invitation to inquire for the kind of relationships, partnerships, and strategies needed to sustain and expand the political creativity that has arisen.

Acknowledgments

I gratefully acknowledge all citizen activists and community members who willingly shared their experience and valuable insights. I would also like to thank all Civil Society at the Crossroads partners and colleagues for their support and comments.

Notes

1. Public letter from Badiou and other French intellectuals denouncing the hypocrisy of the troika which, while claiming to be saving Greece, actually pushes for neoliberal policies serving the interests of the financial system. The letter calls for pan-European resistance and support of the Greek people.
2. A few interviewees referred to "*survivalism*" to point out that the Greek society will either collectively demand a fairer response to the crisis or we will be reduced to the point of fighting against each other to survive.

3. Successive governments have announced new and always final cuts to salaries/pensions or increases to taxes over eight times in the last two years. Any budget planning by households or businesses was rendered useless with each new "measures package".
4. Despite the disconnectedness between grassroots citizen initiatives and formal civil society, an important number of individuals working for trade unions, NGOs and also political parties played an active role in the SSM and other initiatives, making an effort to keep their activist identity separate from their formal affiliation. Activists reported to me that these individuals in some cases supported the formation of new practices through transferring knowledge they have gained in other spaces and that in other cases they have side-tracked the organic processes of movement formation by trying to attract gains for their own organisations.
5. "Collectivities" is used throughout as a direct translation from the Greek word *syllogikotites,* which citizens commonly use to refer to citizen groups and networks that are characterised by horizontal forms of organisation.
6. The geographical focus of the interviews has been Athens both because of methodological reasons and practical limitations. As a metropolis of three million, the city bears the brunt of the human cost of the crisis, witnessing an unprecedented spike in unemployment, homelessness, and destitution. As the centre for national decision-making, Athens has also been the centre of citizens' struggles. Still, it is fair to say that similar initiatives as the ones described here have blossomed in most of the main Greek cities and towns.
7. The SSM – although a child of the Greek condition – took its form through a dialogue with other processes: both of political identification (with the Egyptian demand for "everyone to have a share in shaping politics" and the Spanish cry, "we are not commodities"); and of developing new action repertoires (camping in the streets in Egypt and popular assemblies in Spain).
8. Excerpt from interview with C.M., member of a neighbourhood assembly defining itself as a network of solidarity and resistance.
9. Comment made by C.M. during interview with member of a solidarity network.
10. Direct quotation during interview with K.M., who was one of the initiators of the citizen initiative.
11. Excerpt from interview for the documentary *Solidarity: Antidote to Crisis*, prepared for the online portal www.tvxs.gr.
12. Adapted from AKYA 2005, 13.
13. This section is largely based on and influenced by the work on Unruly Politics by the Participation, Power and Social Change team at the Institute of Development Studies. For more see Khanna (2012) and Khanna et al. (forthcoming) or visit http://www.ids.ac.uk/idsresearch/unruly-politics.
14. *Charatsi* is the popular name for an emergency property tax introduced in 2011. Each citizen has to pay the *charatsi* for the house they live in as an added "contribution to the state" in addition to regular property taxes. In a canny move, the government inserted this tax as an added fee to electricity bills, announcing that electricity provision would be cut from households that refused to pay the extra tax. *Charatsi* was a word used in the Ottoman Empire to describe the taxes imposed by the sultan on his subjects.

References

AKYA (Assembly for the Dissemination of Struggles). 2011. "Introduction." In *Horizontality: Voices of Popular Power in Argentina*, edited by Marina Sitrin. Oakland: AK Press.

Badiou, A., et al. 2012. "Save the Greeks from Their Saviors!Accessed 26 April 2013. http://www.egs.edu/faculty/alain-badiou/articles/save-the-greeks-from-their-saviors/.

Badiou, A. 2005. *Metapolitics*. London: Verso Books.

Douzinas, K. 2011. *Resistance and Philosophy During Crisis: Politics, Ethics and Stasis Syntagma.* Athens: Alexandria.

Economou, M., M. Madianos, C. Theleritis, L. Peppou, and C. Stefanis. 2011. "Increased Suicidality Amid Economic Crisis in Greece." *Lancet* 378 (9801): 1459–60.

Giovanopoulos, C., and D. Mitropoulos. 2011. *Democracy Under Construction: From the Streets to the Square1*. Athens: Asynexeia.

Human Rights Watch. 2012. *Hate on the Streets: Xenophobic Violence in Greece*. Accessed 26 April 2013. http://www.hrw.org/reports/2012/07/10/hate-streets-0.

Hossain, N. 2009. "Rude Accountability in the Unreformed State: Informal Pressures on Frontline Bureaucrats in Bangladesh". IDS Working Paper 319, Brighton: Institute of Development Studies (IDS).

Hossain, N. 2011. "The Moral Economy of the Street: Popular Responses to the Food and Fuel Prices of 2011". Recorded presentation, Accessed 26 April 2013. http://www.ids.ac.uk/go/events/the-moral-economy-of-the-street-popular-responses-to-the-food-and-fuel-prices-of-2011.

Khanna, A. 2012. "Seeing Citizen Action through an Unruly Lens." *Development* 55 (2): 162–72.

Khanna, A., with M. Priyashri, Z. Patterson, M. Pantazidou, and M. Shqerat. Forthcoming. "The Changing Faces of Citizen Action: A Mapping Study Through an 'Unruly' Lens". IDS and Hivos Working Paper. Brighton: HIVOS and IDS.

Kouki, H. 2011. "Short Voyage to the Land of Ourselves." In *Revolt and Crisis in Greece*, edited by A. Vradis and D. Dalakoglou, 167–180. London: AK Press and Occupied London.

Matsaganis, M., and C. Leventi. 2011. "Inequality, Poverty and the Crisis in Greece." Brussels: European Trade Union Institute (ETUIhttp://www.etui.org/Publications2/Policy-Briefs/European-Economic-and-Employment-Policy/Inequality-poverty-and-the-crisis-in-Greece.

Mouzelis, N., and G. Pagoulatos. 2002. Civil Society and Citizenship in Postwar Greece. Paper prepared for the London School of Economics and the Athens University of Economy and Business.

Open Assembly of Citizens of Ag. Paraksevi. 2011. "Experiences and Opinions from the Participation in the Open Assembly of Citizens of Ag.Paraksevi." In *Horizontality: Voices of Popular Power in Argentina*, edited by Marina Sitrin. Oakland: AK Press.

Reason, P. 1994. "Three Approaches to Participative Inquiry." In *Handbook of Qualitative Research*, edited by N. K. Denzin and Y. S, 324–339. Lincoln, London: Sage.

Richardson, L. 1994. "Writing: A Method of Inquiry." In *Handbook of Qualitative Research*, edited by N. K. Denzin and Y. S, 516–529. Lincoln. London: Sage.

Tadros, M. 2012. "Rupturing the Status Quo through Unruly PoliticsAccessed 26 April 2013. http://www.sidint.net/content/rupturing-status-quo-through-unruly-politics-mariz-tadros

Thompson, E. P. 1971. "The Moral Economy of the English Crowd in the Eighteenth Century." *Past and Present* 50 (1): 76–136.

Wainwright, H. 2012. "An 'Excess of Democracy': What Two Generations of Radicals Can Learn From Each Other. Accessed April 26, 2013. http://www.redpepper.org.uk/an-excess-of-democracy/

The author

Maro Pantazidou is a facilitator and researcher working on issues of citizen action, human rights, and social development. She has supported both grassroots and international organisations to produce and use learning resources and to introduce participatory planning and power analysis into their work. She is currently the Learning and Impact Officer for Amnesty International.

Russian civil society: background, current, and future prospects

Charles Buxton and Evgenia Konovalova

The term "civil society" in Russia is often taken to refer to civic organisations and movements created during and after the break-up of the Soviet Union, and is sometimes equated narrowly with "NGOs" – registered non-government, non-commercial, or public organisations. This paper attempts to look at civil society more widely. It considers both registered organisations and more spontaneous/informal civic actions; and follows local experts in challenging the idea that Russian civil society began in 1989–91. The paper considers both recent developments on the ground, and analyses by historians, sociologists, and political scientists that go back to soviet and pre-soviet periods.

Société civile russe : toile de fond, actualité et perspectives futures
L'expression « société civile » en Russie est souvent interprétée comme se référant aux organisations et mouvements civiques créés durant et après le morcellement de l'Union soviétique, et elle est souvent étroitement assimilée aux « ONG » - organisations immatriculées non gouvernementales, non commerciales ou publiques. Cet article tente d'examiner la société civile de manière plus large. Il considère à la fois des organisations immatriculées et des actions civiques plus spontanées/informelles ; et il emboîte le pas à des experts locaux en mettant en question l'idée selon laquelle la société civile russe est née en 1989–91. Cet article se penche à la fois sur les récents progrès sur le terrain et sur les analyses menées par des historiens, des sociologues et des politologues qui remontent aux périodes soviétique et pré-soviétique.

La sociedad civil rusa: antecedentes y prospectivas actuales y futuras
Frecuentemente, en Rusia se utiliza el término "sociedad civil" para referirse a organizaciones y a movimientos cívicos creados durante y después del colapso de la Unión Soviética, a veces equiparándose estrechamente con "ONG"—organizaciones públicas, no gubernamentales, no comerciales y registradas. El presente artículo pretende examinar a la sociedad civil más amplia. En este sentido, revisa tanto las organizaciones registradas como las acciones cívicas, informales y más espontáneas; también da cuenta de la opinión de los expertos locales, quienes cuestionan la idea de que la sociedad civil rusa comenzó en 1989–1991. El artículo examina los acontecimientos que han ocurrido recientemente sobre el terreno, así como aquellos análisis hechos por historiadores, sociólogos y científicos políticos que se remontan a los periódicos soviético y presoviético.

Sociedade civil russa: experiências passadas, perspectivas atuais e futuras
O termo "sociedade civil" na Rússia é frequentemente usado para se referir a organizações e movimentos civis criados durante e após o colapso da União Soviética e às vezes é quase equiparado a "ONGs" – Organizações Não-Governamentais registradas, organizações não-comerciais ou organizações públicas. Este artigo visa examinar a sociedade civil mais amplamente. Ele considera tanto as organizações registradas quanto as ações civis mais espontâneas/informais; e segue especialistas locais que contestam a ideia de que a sociedade civil russa teve início em 1989–91. O artigo considera os desdobramentos recentes na base e análises de historiadores, sociólogos e cientistas políticos que regressam para os períodos soviéticos e pré-soviéticos.

Introduction

The term "civil society" in Russia is often taken to refer to civic organisations and movements created during and after the break-up of the Soviet Union in the early 1990s, and is particularly associated with NGOs. This paper considers both registered organisations and more spontaneous or informal civic actions. Second, it follows local experts in challenging the idea that Russian civil society began in 1989–91. In this way it features not just recent developments on the ground, but also recent analyses by historians, sociologists, and political scientists that take us back into the soviet and pre-soviet periods. A third aspect of the study is that it considers how civil society is developing in a middle-income or post-aid country. With a per capita GDP of US$12,993 and 66th position in the United Nations' Human Development Index in 2011, Russia is indeed a middle-level country on many indices. However, in the 1990s it suddenly transited from being a significant international donor with its own distinctive model of state-led development, to being an external aid and advice recipient.

The 1990s came to an end with Vladimir Putin's ascension to the presidency and the establishment of a very different regime. Putin re-established a top-down order in Russia and has proved to be quite antagonistic both to Western foreign policy and to pro-Western civil society groups (CSOs). However, both capitalism and multiparty democracy continued their uncertain paths in Russia through the last decade. After two terms as president, Putin handed over to Dmitri Medvedev, who pursued a similar line but with a focus on modernisation and a slightly more liberal approach to both foreign investors and home-grown CSOs. And as of March 2012, Russians have Putin as their president again.

Looking to the future, the paper highlights two main trends. First, the continuation of Russia's tradition of civic and political activism, seen most dramatically in the mass demonstrations in Moscow and other cities in the run-up to the parliamentary and presidential elections in late 2011 and early 2012; but also around the country on a range of social and economic themes. Second, the major steps made by CSOs and local government in implementing social partnership, with new funds from the state that both replace and build on the contributions of foreign donors during 1995–2005.

History

The ideology of civil society played a large part in the pro-capitalist revolutions in Eastern European and former Soviet Union countries that heralded their transition periods. In most of them, the new regimes required a re-evaluation or rewriting of history, including both the soviet and pre-soviet period. In Russia this has included efforts both to rehabilitate the authoritarian Tsarist regime, and to identify democratic or liberal trends in the pre-soviet period.

The question of at what point civil society emerged in pre-1917 Russia, and to what extent it possessed the characteristics of civil society in Western European countries, is one of many controversies that have arisen in the reinterpretation of Russian history. Some experts date its appearance to the late eighteenth or early nineteenth century when the first voluntary associations appeared. Others prefer a date around the turn of the twentieth century – when voluntary groups achieved a firmer legal basis and consequently the variety and geographical spread of organisations, their popular support, and relations with government were all more developed.

This paper follows experts at the Civil Society Centre at the Higher School of Economics, Moscow, in dividing the history of civil society in Russia into four main stages: (1) the beginnings of civil society in the late eighteenth century; (2) civil society growth from the time of the 1860s Great Reforms; (3) the Soviet period; and (4) the transition period. Quotations used in the following pages are from the Higher School for Economics (HSE) report (Yakobson et al. 2011, 5–18). A short account of each period is given, with case studies and consideration of how earlier stages have led into subsequent ones (often quite abruptly).

The beginnings of Russian civil society

The first stage (1760–1860) flows out of Catherine the Great's reforms to the Russian estates (*sosolviya*) and was characterised by the creation of public organisations related to science, literature, the arts, leisure, and charitable activities. These included famous and influential associations like the Russian Geographical Society, the Free Economics Society, the Moscow Agricultural Society, the Russian Technical Society, and the Pirogov Association of Russian Doctors. These societies were set up with hopes for friendly cooperation with the Tsarist authorities and in the second half of the nineteenth century their members played a key role lobbying for social and legal reform.

Case study: The Imperial Russian Geographical Society

Founded by prominent scholars Fedor Litke and Karl von Baer in 1846, the Geographical Society became one of the most influential voluntary associations in Russia. The main aim of its initiators was to promote a knowledge and love of Russia. They included many military and government officials and scientists, and when they approached Tsar Nicholas I and his Minister of Internal Affairs for support, they argued that a private society would be able to collect data and popularise findings in a way that the government could not. One of the most important sections of the Geographical Society was its Ethnographic Division. Both Litke and Baer were particularly interested in studies of non-Russian peoples living within the Empire. In 1847, the Society sent out 7,000 questionnaires to all corners of the Empire to collect information on a wide variety of topics: the physical features of the people of Russia; language, dialect, and slang; domestic life including material goods and customs; intellectual, moral, and social aspects of life; and folk legends and memories.

The society opened a Statistics Division that operated very close to government policy at a time when statistics were usually kept secret. It also opened a Political Economy section but then discontinued it. As Joseph Bradley comments, "*a forum outside government in autocratic Russia existed on treacherous terrain*" – but the society "*drew in the public and created civic consciousness*" (Bradley 2009, 105).

Civil society growth under Tsarism

Russia's second stage of civil society development began with the Great Reforms ushered in by Tsar Alexander II in the 1860s. Serfdom was abolished, basic civil rights were established in law, and the first steps taken in the creation of a local government system. CSOs expanded gradually, became more professional, and began to provide educational and health support to vulnerable groups across the country. At the same time industrialisation and urbanisation gathered pace in Russia. The extension of the railway system across Siberia to the Pacific was one of the most dramatic examples of this in the late nineteenth century.

However, the development of capitalist relations in the economy was not mirrored by political changes. The period of reforms gave way to a new period of repression and political stagnation, and the state was challenged by increasingly radical political forces such as the *narodniks* with their "to the people" movement, culminating in the 1905 revolution. Many voluntary associations were radicalised too (including almost all the scientific societies noted above). Significantly, the only law passed in Tsarist time devoted to public organisations was issued by the Senate in the immediate aftermath of the first Russian Revolution, in March 1906. In the next few years, almost 5,000 new organisations, societies and unions were registered. However, once again this reforming, liberalising movement ran into opposition from the state and with the crisis brought about by World War 1, Russia experienced the double revolution (democratic, then socialist) of February and October 1917.

Early Russian charitable associations

Before 1917 Russia lacked a system of social security, and this gap was filled by mutual assistance organisations. This included financial or material support to groups such as widows, orphans, or victims of accidents at work; and "intellectual assistance" to various professional groups (libraries, concerts, literary evenings). Problems tackled included homelessness, abandoned children, crime, and so on. The societies undertook a range of activities familiar to today's NGOs. The Mozhaisk Charitable Association provides an example of the comprehensive approach developed by many of them. Its aims were not simply to feed and care for children, but to teach work skills, find them work, and assist in the development of school education and distribution of textbooks. Later it opened a gymnasium (college) and a sewing workshop for girls, supporting their progress into teacher-training institutions.

In rural areas, the support of *zemstvos* (local authority units set up in the 1860s reform period) was particularly valuable, both as a source of grants for the needy, and for exchange of experience in social care approaches. In urban areas, mutual help associations helped workers and professionals accumulate funds with which to help victims of accidents at work, to pay for medical care or pensions. By the start of the twentieth century, similar associations had been created for writers, artists, lecturers, schoolteachers, doctors, nurses, dentists, merchants, craftsmen, tailors, bakers, and so on – with members paying regular dues (Yakobson et al. 2011).

Civil society in the Soviet period

The Soviet period (1917 to the mid-1980s) is described by the HSE report (Yakobson et al. 2011, 16) as the third stage in Russia's civil society development, notable for the "*nationalization of civil society institutions*". Again, there were stages when popular activity flowered. The Soviet arts, cultural and scientific avant-garde of the 1920s is well known, but less well known are the local movements, peasants' and proletarian organisations that emerged all around the country. For example, every settlement/district (*volost*) had its own Peasant Mutual Society and the Central Bureau of Proletstud catered to the welfare needs of students much as voluntary associations had done before the revolution.

However, in the 1930s this phase gave way to a period of repression and political regimentation – occasioned by the Soviet government's decision to forcibly collectivise agriculture and go for rapid industrialisation. This "*required state-oriented CSOs which were to drive the foundation of socialism*". The voluntary associations created in the 1920s "*offered alternative ways of solving social problems*" but the authorities "*doubted the utility of voluntary movements and the reliability of their participants*" (Jakobson et al. 2011, 16). Thousands were shut down in the 1930s and new associations set up in their stead, as part of the government machine. Eventually only the Red Cross and Children's Fund remained from the original social assistance groups. New mass movements such as the Soviet Committee for Peace, the Union of Atheists, or the Union of Women had an explicitly communist ideology.

It was not until the late 1950s/early-1960s that citizens' organisations of a less politicised type began to re-emerge, encouraged by Kruschev's denunciation of Stalin and the political thaw that followed. Russian analysts have identified about 40 of these, operating mainly within the arts and scientific fields under the patronage of Communist Party bodies and subject to the latter's decisions on policy and personnel matters. By the Brezhnev period, associations were active among groups as varied as war veterans, professional designers, and those involved in child welfare. Civil society now included the dissident movement. Dissidents developed various modes of resistance to the Soviet state – writing and publishing artistic or journalistic critiques of the regime, creating a variety of informal circles and discussion groups, and making statements on political and human rights issues that brought down considerable persecution on themselves. But they had many sympathisers and considerable impact on the political atmosphere in the country and its reputation abroad. The dissident movement included not only Western-leaning liberals, but also strident nationalists and religious activists from many of the constituent republics of the USSR. The HSE study concluded that at grassroots level the shoots of civil society had begun to emerge: that is, the commitment of citizens to self-organisation within socially oriented activities.

Transition period to date

As noted above, civil society played a huge role in dissident ideology in Eastern Europe and the USSR in the 1980s. When the communist regimes collapsed one after the other it seemed to many people that an entirely new society was being born. Many different analyses of newly-created civil society in the region were built around this view.

Twenty years later, and with the benefit of new historical research, most experts in the region see things differently. The current stage of civil society development in Russia is a fourth stage, starting in the mid-1980s and continuing to the present day. The movement for *perestroika* and *glasnost* led by Gorbachev was designed to solve the USSR's pressing economic crisis (caused by the arms race and economic competition with the West) and to shore up the legitimacy of one-party rule. But it led instead to the collapse of the communist system. Many of the most active civil society sectors today can trace their origins to the 1980s – not just the human rights groups, but also the environmental movement with its active networks among young people and in the regions. The adoption of a law on public associations in the late soviet period, supported by subsequent Russian Federation laws regulating public and charitable activity, opened the door to CSO registration for all-comers.

The role of foreign donors was very significant in the 1990s, enabling the transfer of a Western (primarily American) model of civil society to Russia. At the time the whole FSU region was in deep recession, with a 25–50 per cent drop in GDP and production, political and intellectual links between the 15 former socialist republics in disarray. The HSE study (Yakobson et al. 2011, 17) describes the role of the state during the 1990s as "*benevolent non-interference*"; state budgetary support for CSOs was "*insignificant*". However, under Yeltsin's successor Putin, the environment for civil society changed significantly. On the one hand, Putin accelerated a process initiated under Yeltsin – government financing of the sector (mainly via contracting out social services to CSOs) – and set up a national structure of Public Councils to dialogue with and co-opt the sector. On the other hand, in 2006 he introduced regulations limiting the influence of foreign donors. The study's authors call this policy "*import substitution*"; the replacement of foreign models and funding by national programmes, self-organisation and local philanthropy. This paper will briefly indicate how successful this process (which is still quite new) has been.

Russian civil society today

General features of the sector

After two decades of transition, analysts of current developments in Russia are beginning to gain a balanced view of the civil society sector. *Russian Civil Society: A Critical Assessment* is one of the most thorough recent books on the sector (Evans, Henry, and Sundstrom 2010). In a concluding chapter on "Tensions and Trajectories" the authors note the black-or-white nature of many analyses of civil society in Russia, while more attention to the detail shows that elements of continuity and change, tradition and innovation, exist alongside one another. Here we quote a small selection of the features that this important analysis identifies:

- Civil society activists today are "*a strong minority of citizens*" who deserve more support
- Informal networks are important for civil society, especially in rural areas because they include a large membership and their ability and readiness to provide vital daily services, plus often good links with government
- Foreign funding has had a positive effect in many areas (for example, it helped open up dialogue on many issues like feminism, domestic violence and others)
- Mafia-type groupings have had a powerful and negative effect at all levels in Russia – even "*co-opting the role of civil society*"
- The millionaires or oligarchs that emerged during the 1990s preferred not to work through formal or wider business associations; hence they contributed little to civil society development (305–33).

The authors conclude that a "*more Russian form*" of civil society is emerging – a "*really existing*" civil society (319). The question is how to develop it further, from local to national level.

The HSE team recently produced a similarly balanced view of civil society development in Russia, using the CIVICUS model with its five key dimensions of civil society, based on the results of a nationwide sociological survey and a range of individual and group interviews. A brief summary of the conclusions for each dimension is given below:

- Civic engagement: the level of public activity in stable periods is moderate, but CSOs have shown they can mobilise quickly when the external situation demands
- Level of organisation: Russian civil society is still in a difficult period of organisational development
- Practice of values: CSO members and activists refer to non-violence, tolerance and internal democracy as being among the most important for the sector

- Perception of impact: the general image of civil society is not equal from internal and external points of view. CSOs themselves rate their social and political impact as higher than the scores given by external experts
- External environment: the majority of the population do not approve of corruption, tax evasion, and so on. These positive social attitudes could potentially act as a catalyst for further civil society development (Yakobson et al. 2011, 19–36).

The legal and political environment for civil society

The sources quoted above show a complex picture, with significant advances made and yet many problems remaining. This view is echoed by experts in NGO law working in the Siberian city of Novosibirsk, who pointed to the bias in many Western media reports. These, they argue, either select data from Moscow (not representative of Russia as a whole) or from particular controversial sectors like human rights and election monitoring (rather than the social or economic sectors) (Javeline and Lindemann-Komarova 2010). "*Overly generous assessments of Russian democracy*" in the Yeltsin period have, they claim, given way to misleadingly negative commentary on the Putin period (172). They cite three key features of the external environment for NGOs where a more careful analysis would be useful.

Registration of NGOs: Russian Federation Law No.18-FZ was adopted in 2006 and was heralded as an "*unprecedented assault on the work of human rights groups*" (as quoted on the Human Rights Watch website, December 27, 2010). Freedom House noted "*the government intensified crackdown on NGOs, particularly those receiving foreign funding*" (quoted in Javeline and Lindemann-Komarova 2010, 173). Javeline and Lindemann-Komarova note, by contrast, that legal requirements for NGOs in Russia, particularly financial reporting, have always been onerous and surveys carried out in 2007 indicated that Law No. 18 was no more burdensome for human rights and advocacy groups than other NGOs. On the other hand, four out of five NGOs in 2007 did not supply reports in the required format by the official deadline – which left them vulnerable to forced liquidation. A survey of NGO leaders in the same year asked them to name the main sources of pressure on their organisation. "*Not enough money, material resources*" topped the list with 59.1 per cent, while government pressure came very low (2.9 per cent).[1]

Local self-government: At the same time as the rules for NGO registration were changed, another law came out – On General Organisational Principles of Local Self-Government in the Russian Federation. This law, Law No.131-FZ, boosts citizens' right to participate in local self-government bodies. Javeline and Lindemann-Komarova call this a "*huge stride for Russia in recognizing the value of citizen participation in governance*" (2010, 175). At the same time they note scepticism remains because many procedures (like public hearings and referenda) can easily be manipulated. Much depends on the experience and influence of individual civil society leaders and organisations.

Consultative mechanisms: Set up by Putin in 2004, the national Public Chamber has 126 members, selected in equal numbers by the President, public organisations, and Russia's regions. The aim was to develop the space for civil society and inter-sectoral dialogue. Gradually, this model has been extended across Russia. Critics said they would prove to be mere "*window-dressing to legitimize the government's increasingly authoritarian policies*", but at local level many NGO supporters have proved willing to give them a try. A survey in Novosibirsk found that 60 per cent of activists would participate if asked, while 16 per cent would not (Radio Free Europe, 18 February 2005, and a survey by Novosibirsk SCSIC; both quoted in Javeline and Lindemann-Komarova 2010, 180–81).

Regional case studies: experience from Urals and Siberia

In the second half of this paper, several examples of civil society development and activism are taken from the Urals Mountains region and Siberia.[2] These regions were chosen to provide a chance to look beyond Moscow and St Petersburg into the remoter parts of Russia, colonised in stages from as early as the sixteenth century but still facing huge development challenges. In fact, the authors of this paper have a special connection with the Urals. One of us was born in the Komi Republic (northwest Urals); the other visited the region several times from the mid-1990s within a variety of development programmes and research initiatives. In the 1990s, the handover of a variety of services from factories and enterprises to local government had only just taken place, and international assistance in setting up new health, social services, and education departments, plus the beginnings of CSO involvement in social programmes, were much appreciated by local partner organisations in Russia.

Ten years later, visits undertaken for this paper went eastwards, taking in a central belt of Siberia from Novosibirsk in Western Siberia to Ulan Ude, capital of the Republic of Buryatia, east of Lake Baikal. The rest of this paper is based on information collected during these visits, and illustrates the general points about civil society made above with examples from two major regions in the borderlands between Europe and Asia. It considers the current priorities of "organised civil society" (i.e., registered NGOs) and the wider mass of civil/political activists campaigning on the key current issues in Russia. This is necessarily a brief description and the reader should not draw extensive conclusions from the examples given.

Civil society and local government in Perm

People both in and far beyond Perm regard this as one of the friendliest, most innovative regions in Russia for civil society. A system of "social orders" – social services contracts let out by local government to NGOs – has been running for over ten years and NGO representatives play a significant role in designing programmes, approving contracts and monitoring the process. There is an active Public Chamber bringing together the government and civil society sectors.

Nonetheless, in Perm region as a whole, many serious economic and social problems remain. While the region has some advanced industries (including chemicals, mining, and machine building), a recent UN report on human development noted a continuing decrease in the population, low life expectancy among men, and problems of half-deserted villages and polluted mono-industrial towns (UNDP 2010, 58–62, 94–96). This is less evident in the regional capital, Perm city, where there are many signs of increasing prosperity and a concern for good governance. The efforts of local government to work closely with civil society are shown, for example, by the opening of some 20 new community centres with rent-free space for NGO offices and a hall for local events. Apparently, the idea came from exchange visits by local officials to the UK.

Another example of local inter-sectoral collaboration is an independent NGO, the Grani Analytical Centre, that specialises in governance issues and has begun working for a variety of local and federal agencies, for example the Ministry of Economic Development, to evaluate efficiency and accountability in government. One study is monitoring the work of the Federal Anti-Monopoly Service; another piece of work is on land registration services in five regions of Russia; a third was on how well citizens participation works at municipal level.

Putin's first period in government became well known for his re-centralising policies: the new President argued that this was essential to stop Russia from falling apart – to stand up to regionalist tendencies and in particular to reverse the military failures suffered in the First

Chechen War. However, during the same decade some initial steps were made in delegating power to local government; for example, like some other Russian cities, Perm experimented with directly elected mayors. In summer 2012 it seemed likely that a newly re-elected Putin might reverse his policy of appointing the governors in Russia's super-regions. The new law on local self-government (mentioned above) is another important step, but one that will require civil society and political pressure to really work.

Perm is also notable for its attempts to promote the liberal and human rights agenda. One example is its support for the biennial "Pilorama" festival – held two hours' drive north of Perm in one of Russia's most notorious forced labour camps, organised by a coalition of human rights activists and organisations including the national human rights organisation Memorial. In summer 2011, the festival featured a wide range of celebrities, from *perestroika*-era politicians to rock stars, as well as providing a forum for political debate in which civic groups as well as political parties played an active role. Another example is the efforts of the local authorities to promote Perm as a centre of progressive art and culture – the Perm "cultural breakthrough" as they call it. There is a modern art museum in the old Kama River Station, a highly visible city graffiti programme, a controversial new city emblem, a multicultural, multidisciplinary White Nights festival, and a commitment to maintain and diversify Perm's musical traditions. The result of this almost everyone in Perm seems to have a view on modern art – for or against – and this is becoming a significant factor in the city's perception of itself, in its civic consciousness, and the promotion of a more open, liberal approach.

Government funding for civil society: the Siberian experience

The NGO resource centre in Perm is closely involved in new government funding initiatives for the civil society sector – as are many experienced NGOs and individual experts across the Urals and Siberia. A special role in this effort is played by the Siberian Civic Initiatives Support Centre (SCISC) in Novosibirsk. With over three million people, Novosibirsk is the third largest city in the Russian Federation. Situated on the Ob River, this is a major centre for machine building and the automobile industry in particular.

SCISC was one of the first generation of NGO resource centres, set up with US funding in the mid-1990s and serving a network of similar organisations in 11 administrative regions across Siberia. In 1996, the Centre received US$ 600,000 from USAID to organise one of the first small grant competitions in Russia. It used this money to open discussions with the Novosibirsk authorities about government funding for NGOs – eventually gaining an agreement to match the US contribution. Novosibirsk Oblast and the city of Novosibirsk awarded grants to NGOs totalling 18 million rubles in 2007, and the funding pool has now topped 35 million rubles (US$ 1.1 million). Public consultative committees, neighbourhood councils, and local volunteer schemes have been developed as part of the same programme.

Alongside the national Public Chamber (see above) there are now over 50 similar bodies across Russia. New resources provided by central government enable them to make contacts and study processes and results across the whole civil sphere. A report commissioned in 2010 by the Public Chamber in the Siberian federal region (*okrug*) on the number of registered NGOs (overall, the number of NGOs dropped slightly because of tighter government requirements for registration and reporting, as mentioned above), analysed them by type and sector, and devoted other sections to political parties, youth movements, trade unions, and veterans associations. The report shows that civil society development in the region is quite uneven. For example, anti-corruption committees were reported to be active in the Siberian cities of Omsk, Krasnoyarsk, and Irkutsk. Budget hearings and citizen's audits took place in several areas – for example in Altai region in the discussion of a new federal law on the police. By contrast, there is still no chamber in the Autonomous Republic of Buryatia.

According to the umbrella NGO, Association for Social Information (ASI), public chambers should be seen as a bridge between the government and the people (Topoleva 2011). They are "invited spaces" with a mainly top-down function – but they can be used, for example, to lobby environmental issues with private sector companies. An example of this is in Siberia, where quite suddenly a broad coalition emerged to oppose a new pipeline being laid on the northern shore of Lake Baikal. Victory in the campaign against proposals by Russian oil company Yukos and its successor Transneft, was due not just to the citizens and environmental groups who organised meetings and collected thousands of signatures to oppose the plan, but also to local government leaders who decided to support them. In April 2006 Putin responded to local pressure by ordering the re-routing of the pipeline.

Moscow and the regions

In all of these developments, a clear gap can be seen between advanced and less advanced elements in civil society and government. The rural areas lag behind the cities, the remote regions lag behind the industrial centres. The character of political opposition to the regime is quite different in the regions. As the demonstrations of autumn 2011 and spring 2012 showed, present-day dissidents in Moscow and St Petersburg tend to be middle-class, liberal, and western-leaning. However, in the Urals and Siberia, many of the most strident activists are from the communist and nationalist camps.

This is not to say that pro-democracy forces should be discounted. Thus in Novosibirsk we were told by NGO Golos ("voice") that election monitoring was actively and efficiently organised across the nine regions of Siberia and the Far East which they supervised. Activists from a new radical opposition grouping in the city were enthusiastic about the first mass movement on this scale since 1993 and albeit limited success in placing demands for greater election process transparency on government – for example, video monitors in all election rooms. In the wider civil sphere, a coalition called Regions was recently set up to campaign for more attention to organisations and issues outside Moscow and St Petersburg. NGOs in Siberia clearly feel excluded from the inner circle of Moscow-based NGOs; representatives of the coalition now sit on various committees to ensure more equal access to government decision-making and funds.

Civic activism and social movements

During visits to the Urals and Siberia in 2011–12, we spoke to civic activists working on a wide range of issues. At the Pilorama Festival, activists debated democracy and privatisation, issues on which there was little general agreement. A visit to a youth ecological camp in 2012 showed that environmental groups are still active but have difficulty institutionalising some of their gains; for example, despite a huge amount of lobbying in the 1990s, it has been impossible to gain a satisfactory place for environmental studies in the school curriculum.

In Novosibirsk, Irkutsk, and Ulan Ude, we interviewed social and environmental sector activists, including representatives of informal groups linked with the Institute for Collective Action. One of these, an activist in the pensioners' movement, stated that he had been involved in over 250 court hearings in the past year. The restrictions on street protests are such that it has become necessary to apply officially for any picket or street demonstration. The main actions he described included a campaign to force the city authorities to adopt a more flexible system of free/subsidised travel for pensioners on public transport; a campaign to stop the cutting down of trees to make way for a new hotel in a local park;

and an anti-corruption action against the management of a major cement factory in the city. He and his friends, many of them retired people, had succeeded in making contact with a wider protest constituency during the upsurge of political activity the previous winter. On one occasion they had assembled almost 7,000 people for a demonstration around the elections. But within a few weeks the younger people disappeared again and it was back to the "*regulars*".

In Irkutsk, the editor of an independent local newspaper *Narodny Kontrol* (People's Control) described efforts to highlight and uncover local corruption and malpractice. He sees his newspaper's main goal as being to mobilise public opinion and influence government. The paper had printed an extra 200,000 copies of its latest issue for circulation around Irkutsk city and oblast. "*If I can help people to get a sense of what is going on – in their own head, in their courtyard, in the city as a whole – then I have achieved my aim*", he told us.

The report of the Siberian Public Chamber has some interesting comments on the topic of protest. Analysing recent public opinion surveys on citizens' dissatisfaction with the government, it notes that the main causes for dissatisfaction are price rises, unemployment, and housing problems. Some 32 per cent of the population defined itself as "poor" and only 43 per cent described the current economic situation as "positive". Some 900 protest actions were recorded in the Siberian region in 2010, with 75,000 participants in total, the majority organised by political parties. But the report also notes examples of more spontaneous protests. One of these was the campaign for kindergarten provision. In meetings across Siberia, it was interesting to hear first-hand accounts of a fightback 20 years after the mass closure of childcare institutions during the 1990s. A childcare activist in Irkutsk explained that one of the consequences of the federal government's pro-natalist policies is an increase in the birth rate locally, with some 20,000 families now on the waiting list for nursery places. A new movement led by young mothers is called 38mama.ru; it is very active and uses a variety of informal means including new/social media to gather support.

Problems of coordination on social and political issues

If we accept that civil society is a kind of space, we will probably agree that it does not *belong* to any single group – political activists, local problem-solving organisations, expert or professional associations, etc. But the question remains, to what extent civil society can raise issues that challenge society as a whole? By using Russia's anti-globalisation movement as an example, we can examine to what extent a social movement of this kind is able to unite activists or present what Gramsci would call a counter-hegemonic project to the main line of government and business sector thinking.

The Russian anti-globalisation movement

One of the campaigns that could potentially unite activists promoting progressive or radical causes is the anti-globalisation movement. Russian anti-globalism has a long pedigree, with roots in nationalist, Slavophile, and conservative thinkers who up to the Bolshevik revolution promoted Russia as a special civilisational centre, opposing westernisation and what they saw as its evils – immorality and commercialisation of life. According to a leading spokesman for Russian anti-globalism – or as he prefers to call it, alter-globalism – the main enemy is neoliberalism and world domination by the USA (Buzgalin 2006). The alternative is provided by grassroots social movements including workers' organisations, pensioners and housing movements, environmentalists, anti-debt campaigners, and so on. Some of these social movements have a local focus and an everyday social or political character, while others focus on higher-level national or international policy issues and opportunities for action; including protests around the World Trade Organisation (WTO) and participation in World Social Forum events.

Anti-global mass actions in Russia began in November 2001 with the country's first street protest against membership of WTO. In the following year, similar demonstrations were held when George Bush visited Russia, and in 2006 during the G8 meeting in St Petersburg. These protests were not as large as demonstrations held in London, Paris, or Prague; nor did they escalate into Gandhi-type non-violent action.[3] Nonetheless, they were very new for post-1991 Russia – anti-capitalist actions taking a lead from global civil society. Indeed, joining WTO carries many risks for Russia. The manufacturing sector has been in recession since the early 1990s, and faces huge competition from the West and China; while agriculture will always require special government support because of Russia's harsh northern climate.

Siberian social forums: One of the high points of social movement activism in Russia in recent years was the series of social forum meetings held during the 2000s. Five forums were held in Siberia in 2003–08, in Barnaul, Novosibirsk, Tomsk, and Irkutsk; and two national social forums were organised in Moscow. An account of one of the forums held in Novosibirsk shows the variety of participants and the difficulty of agreeing aims. Some 250 participants attended this event from 13 regions. Izhevsk and Perm were described as examples of active social movement coordination; in the latter, workers' hostel residents had organised a successful campaign to privatise their living accommodation. Another important action discussed was the strike by car workers in the city of Togliatti. An ecological section at the forum attracted 50 people. In its modern version, Russian anti-globalism is supported not just by libertarians and the new left but also by nationalist forces. This is a similar situation to that faced by the pro-democracy movement. It means that the movement is split outside parliament as well as having limited support from the political parties inside parliament. A further challenge, according to analysts at the Institute of Globalisation and Social Movements (Kagarlitsky 2006),[4] is that the Russian movement lacks internationalism and has been unable to develop strong links with other global campaigns (such as the Occupy movement).

Conclusion: results and future prospects

In December 2011, Russia was admitted to WTO (18 years after first applying) and all that remained was for the Duma to ratify the agreement. In March 2012, Putin returned as President for a new seven-year term. Thus, for the alter-globalisation movement as for other dissident forces, a new period of struggle lies ahead to win a greater degree of social and economic justice and a more democratic society in Russia. On the other hand, Putin has promised to prioritise social issues and CSOs are well placed to work for positive results in this area. This is a kind of crossroads for Russian civil society: will activists and organisations collaborate or conflict with government? It is likely that many will opt for the first of these strategies, so they face the challenge of how to resist incorporation and maintain their own agenda during the negotiations.

Both the democracy and the alter-globalisation movements face the challenge of how to coordinate their activities, combine different viewpoints, and communicate more effectively with the general public. The activists involved in social issues at local level have put energy and ideas into setting up grants contests, ensuring transparency in awarding contracts. Now they need to focus on the development, implementation, and monitoring of longer term programmes, like anti-corruption, anti-racism, or the protection of minority rights. With all these challenges and changes, another priority is research and reflection on civil society development. Here too there is work in progress and this study has attempted to illustrate and learn from it.

Acknowledgements

Special thanks are due to Moscow-based colleagues in the Association for Social Information, the Civil Society Studies Centre at the Higher School of Economics, and the Institute for Collective Action for ideas, contacts, and analyses that this study has drawn on; also to civil society activists and friends in the Urals and Siberia regions who provided valuable insights into civil society development and activism.

Notes

1. The surveys referred to were carried out by the independent (US-funded) organization ICNL. The same article notes that in 2007, 2,600 Russian NGOs went into involuntary liquidation (out of 675,000 NGOs registered around the country that year). To register a NGO costs 2,000 rubles (around US$ 65) plus legal advice – the latter is available free from NGO resource centres and a number of websites.
2. The Urals Mountains are often referred to as the boundary between the European and Asian continents.
3. The repertoire of protest action in Russia is increasingly colourful, especially anarchist and punk actions (e.g. Pussy Riot) aimed at ridiculing establishment positions.
4. For more on national and international solidarity campaigns, see www.ikd.ru and www.igso.ru.

References

Bradley, J. 2009. *Voluntary Associations in Tsarist Russia: Science, Patriotism, and Civil Society*. Cambridge, MA: Harvard University Press.

Buzgalin, A. 2006. «Антиглобализм»: в поисках позитивной альтернативы новой империи. Accessed April 26, 2013. http://www.alternativy.ru/ru/node/76.

Evans Jr., A., L. Henry, and L. M. Sundstrom. 2010. *Russian Civil Society: A Critical Assessment*, edited by A. EvansJr., L. Henry and L. M. Sundstrom. Armonk, NY: ME Sharpe.

HSE. 2011. *Справится ли государство в одиночку? О роли НКО в решении социальных проблем. Российские общественные организации и решение социальных проблем: ретроспектива*. Moscow: HSE.

Javeline, D., and S. Lindemann-Komarova. 2010. “A Balanced Assessment of Russian Civil Society.” *Journal of International Affairs* 63 (2): 171–88.

Kagarlitsky, B. 2006. «Петербургская повесть. Взгляд.» Accessed April 26, 2013. http://www.duel.ru/200631/?31_3_1.

Topoleva, E. 2011. So Much and So Little: Progress and Prospects in Health, Welfare and the Role of NGOs in Russia. presentation at the Bearr Trust 20th Anniversary Conference, London: November 18 http://www.bearr.org/en/event/conferenceNovember2011.

UNDP. 2010. *Regional Human Development Report for Perm krai*. Moscow: UNDP Russia Country Office.

Yakobson, L., I. Mersianova, O. Kononykhina, V. Benevolenski, E. Pamfilova, L. Proskuryakova, and A. Tumanova. 2011. *Civil Society in Modernizing Russia*. Moscow: Centre for Study of Civil Society and the Non-Profit Sector at Higher School of Economics, National Research University.

The authors

Charles Buxton (corresponding author) is a Senior Capacity Building Specialist for the International NGO Training and Research Centre (INTRAC), and is based in Bishkek, Kyrgyzstan, as Programme Manager for the Central Asia region.

Evgenia Konovalova is a Trainee Researcher at the Centre for Civil Society Studies and the Non-profit Sector at the HSE. She is involved in research work on civil society and the third sector in Russia.

Civil societies at crossroads: lessons and implications

Rajesh Tandon and L. David Brown

The papers in this Special Issue are drawn from a larger set of cases of citizen activism and civil society evolution developed by the "Civil Society at Crossroads?" initiative. What can be learnt from comparative analysis across these stories? In this concluding paper we seek to identify lessons that emerged from comparative analysis across cases and to explore their implications for civil society practitioners and policymakers.

Des sociétés civiles à la croisée des chemins : enseignements et implications
Les documents figurant dans ce Numéro spécial sont tirés d'un ensemble plus large de cas d'activisme citoyen et d'évolution de la société civile préparé par l'initiative « Civil Society at Crossroads? ». Quels enseignements peut-on tirer d'une analyse comparative de ces différentes histoires ? Dans ce document de conclusion nous cherchons à identifier les enseignements qui se sont dégagés d'une analyse comparative de tous les cas et à examiner leurs implications pour les praticiens et les décideurs de la société civile.

Las sociedades civiles en la encrucijada
Los artículos para este Número Especial se seleccionaron de un conjunto más amplio de casos de activismo ciudadano y de evolución de la sociedad civil, a partir de la iniciativa "¿Se encuentra la sociedad civil en la encrucijada?" ¿Qué hallazgos pueden generarse a partir del análisis comparativo de estas historias? En este artículo final, los autores pretenden identificar los aprendizajes surgidos del análisis comparativo de varios casos y examinar sus implicaciones para los operadores de la sociedad civil y para los formuladores de políticas.

Sociedades civis na encruzilhada: lições e implicações
Os artigos desta Edição Especial foram retirados de um conjunto mais amplo de casos de ativismo de cidadãos e evolução da sociedade civil desenvolvido pela iniciativa "Sociedade Civil na Encruzilhada?". O que pode ser aprendido com uma análise comparativa entre estas histórias? Neste texto de conclusão, buscamos identificar lições que surgiram da análise comparativa entre os casos e explorar suas implicações para profissionais da sociedade civil e formuladores de políticas.

Emerging lessons

We have identified eight lessons in our comparative analysis. These patterns emerged across many cases, but we illustrate them here with the cases described in this Special Issue.

1. Citizen protests reflect the gap between their expectations and the performance of public authorities

During the past two decades, rapid economic growth and waves of democratisation have encouraged citizens to develop higher expectations for living standards and government performance. International agencies increased their support to poverty eradication and democratic governance. NGO programmes in developing countries promoted improved livelihoods and basic services to the poor. "Rights-based" approaches organised marginalised groups to hold duty-bearers accountable (Beauclerk, Pratt, and Judge 2011). Cooperation with market institutions through programmes such as microfinance engaged bankers and finance executives to deliver services to the poor. Information and communications technology vastly increased citizen awareness of economic and political conditions in other countries, including the industrialised world. All these factors raised citizen expectations about economic benefits and political freedoms (Knight, Chigudu, and Tandon 2002). Rights, voice, and dignity became part of everyday expectations of citizens around the world – and in many cases those expectations have not been met.

Many cases of citizen activism protested against public authorities. The Chilean students' movement began as a protest against the "selling-off" of a private university that symbolised government failure to deliver universal access to quality education. The Traders Association in Uganda protested retroactive enforcement of ruinous new financial policies and their inability to get any response from the Finance Ministry. Greek citizens went to the streets to challenge the "austerity" measures and politicians who were more accountable to the European Union (EU) and banks than to their citizens (Badiou et al. 2012). Cambodian farmers protested government authorities' connivance with private contractors in land grabs (Ojendal and Lilja 2006).

How can these new citizen movements be understood? Several underlying forces helped produce the tension between citizens and the public authorities. First, the rapid expansion of democracy and market-oriented economic policies in the early 1990s created widespread citizen expectations for improved quality of life, but stories from India, Greece, South Africa, Chile, and UK suggest that those expectations did not materialise for most citizens. Second, concentration of power and wealth in the hands of a few elites increased dramatically in many countries, including political leaders linked to powerful business interests in some. Citizen protests in Greece, India, and the UK challenged takeovers of democratic institutions by such elites. Third, in many countries elected leaders failed to hear or respond to citizen concerns, as in Uganda, India, Cambodia, Chile, and Russia. In several instances, public regulatory authorities appeared to have been "captured" by private interests. Existing channels did not enable public accountability. Fourth, new information and communication technologies (ICT) have enormously increased information flow, awareness, and solidarity amongst the citizens, especially about abuse by the elite (Sitapati 2011). Enhanced communications informed and enabled citizens' movements in Argentina, Chile, Greece, and Russia.

The sudden nature of many citizen eruptions has taken public authorities by surprise, though they were often preceded by years of frustration and anger. Existing formal associations – political parties, trade unions, NGOs – did not provide adequate voice, so citizen "eruptions" suggest disconnects within civil society as well (CIVICUS 2011). But there may be distant complementarities underlying this proximate isolation; conscientisation and mobilisation by civil society organisations has helped build citizen awareness and associations. These organisations include citizens from across many divides – urban and rural, lower and middle classes, men and women, young and old, North and South. The movements of disillusioned citizens emerged in situations of relative deprivation (as in Cambodia, India, and Uganda) as well as in relative prosperity (as in Chile, Russia, Greece, and the UK).

2. New civil society actors are organised for expression, inclusion, participation and innovation

Over the past two decades increased funding to formally organised NGOs, including INGOs and their southern partners, has created pressure for enhanced organisational efficiency and effectiveness. Interventions, improved programme planning, implementation, and monitoring; experts restructured financial management and governance; organisation development became popular (Lewis and Kanji 2009). The models of organisational effectiveness were largely borrowed from the corporate sector. Results-based management, log frames, and "value for money" dominated leadership attention. These approaches can be important assets for large-scale service delivery. But their emphasis on efficiency and finance may also distract civil society organisations from values of local empowerment, capacity building, and systemic change that are not easily measured in financial terms (van Til 2000). Many civil society organisations with external funding have become more formal and professionalised as well as larger and more complex.

Many of these cases of citizen activism displayed quite different forms of organisation from those that had become typical of many NGOs. The Chilean student movement was notorious for creative protests and their use of social media; the movement drew on leadership from many sources, although the media became obsessed with a leader who also happened to be a beautiful young woman. The Occupy movements in North America and Europe defied traditional organising logic, as citizens of all persuasions – young and old, students and workers, housewives and professionals – gathered to express concerns of "the 99%". Occupy London was criticised for not adopting explicit goals, though its members agreed that raising awareness and encouraging experiments in "*being the change*" were more important. The educational wing of the movement rejected the central concern with specialised roles and expertise in its emphasis on "*everyone a learner and everyone a teacher*" (Howard and Pratt-Boyden, 2013, this issue). In the Netherlands, the Makassar community relied largely on local volunteers to organise innovative community activities, though they also became acutely aware of the importance of "professionalization" as a way to manage tensions among community interests.

Some initiatives revived cultural forms to mobilise citizens and enable conversation. Street theatres, arts, mimicry, songs, poetry, music, and a wide spectrum of communication methods enabled discourse on the issues and entertained the public in Chile, Greece, and the UK. Methods of individual, home-based, yet collective actions have also gained considerable currency around the world. Internet petitions, lighting candles, or pot-banging at pre-determined times do not require physical proximity but do express collective solidarity and demonstrate widespread support.

Most movements used non-violent and peaceful methods, though violence did occur in some cases, as in Chile and Greece. Common actions included hunger strikes, sit-ins, occupations of public spaces, acts of civil disobedience, and defiance of authorities. These citizen actions have gained currency from Greece to Russia, Ireland to UK, Malawi to South Africa, Chile to Argentina, and Cambodia to Indonesia. Methods of mobilisation, organisation, and protest popularised by Mahatma Gandhi are combined with modern forms of social media and internet communications.

What can explain such alternative ways of organising civil society? Many formal institutions have become alienating hierarchies that seek to improve efficiency by marginalising human spontaneity (Papadakis 2006). More horizontal and spontaneous organisations have wide appeal. It is also true that young people often reject existing approaches, and the new social media has supported their interest in experimentation. A third factor is that previous civil society efforts to promote self-help and empowerment have led many citizens to question established procedures, mechanisms, and norms (Kilby 2011). As new ways of organising civil society are invented, we expect hybrid forms will combine the formal organisational approaches of NGOs with the innovations from new citizen collectives. This integration will be especially relevant as citizens' movements seek sustainable solutions that require durable and large-scale organisational mechanisms.

3. Partnerships with old and new media expand and reshape civil society outreach and impacts

Historically civil society actors have often been tentative about engaging with the media. In many countries the old media (newspapers, radio, and television) were largely controlled by governments, so civil society access and advocacy was restricted. International media were more accessible, especially to international NGOs (INGOs). Local civil society actors often lacked capacity to engage the media and many focused on work with local constituencies in remote areas, so their work was of little interest to the existing media. Many civil society actors feared unfavourable publicity. Civil society partnerships with media were rare. The new media available through the internet and mobile phones has only recently emerged as an important factor.

Citizen activists have found new media very user friendly: widespread access to mobile phones and the internet has facilitated horizontal communication, leadership, and solidarity. Demonstrations in Chile, Greece, the UK, and Uganda have made extensive use of mobile phones and internet access to launch and coordinate their activities. In Chile, cell phone pictures of police provocations were used to debunk accusations that the student movement was the sole cause of violence. The Chilean student leaders used mobile phones to catalyse pot-banging demonstrations of widespread support for their campaigns. NGOs have also begun to use new social media for public information and education, resource generation, membership engagement, and communication with policymakers. But the new media can also confuse and distort issues with counter-voices; blogs can misinform as well as inform. The open access of social media is both its strength and its limitation.

Citizen initiatives have also increasingly used the old media in their campaigns. Those media have become far more accessible in many societies with the rise of independent TV channels, radio stations, and local language newspapers. Civil society has also become more interesting for the old media. Many newspapers and TV channels devote regular attention to civil society. Thus, the Greek protests and Occupy movements became globally known through coverage in old media. The Chilean student protests and HIV/AIDS campaigns of South Africa gained widespread visibility through radio, television, and newspaper stories.

Several shifts in the past two decades help explain these new partnerships. The expansion of independent TV, radio, and newspaper outlets created more competition and options, especially where official media prevented challenges to the government. The rise of independent privately-owned media offers alternative outlets, but private ownership can also lead to media dominated by private interests. The media in Chile, for example, gave more attention to a beautiful student leader and a few violent masked anarchists than to the issues raised by thousands of peaceful protesters. Journalists in Argentina supported the cause of sexual minorities but their editors and publishers did not. The global media continued to press for austerity measures in Greece rather than for alleviating the sufferings of its citizens. So changes in old media ownership may diversify its coverage and sympathies enough to make more partnerships with civil society possible – but challenges remain.

4. Contracting resource bases reshape civil society roles and relations with government and business

The role of civil society organisations was expanded in many countries in the last half of the twentieth century. Many industrialised countries contracted out public services to civil society organisations (Salamon 1995), particularly in European countries like the Netherlands and the UK. In developing countries, many aid agencies turned to INGOs and their local civil society partners to catalyse sustainable development (Lewis and Kanji 2009). As a result, civil society actors became accustomed to ready access to financial resources.

The shifting global order is altering access to those resources. In many developing countries, Official Development Assistance (ODA) declined in the past decade. Countries like India and South Africa "graduated" to middle-income status that implies less need for aid. Aid to most countries in Latin America began to dry up a decade ago. Donor attention has shifted to the poorest countries in Africa and Asia and reduced support to countries like Uganda and Cambodia. As flexible funding for civil society organisations has become increasingly scarce, many seek new ways to mobilise resources. Some have engaged in microfinance and social enterprise activities that have pressed them to adopt market-based principles. Others are exploring contracts with governments to deliver public services, as in India, South Africa, Argentina, and Uganda. Relying on government funding compels civil society organisations to work within government frameworks and to compete with for-profit providers. Still others seek resources from the growth of private philanthropy as developing countries produce increased wealth. Emerging philanthropists and corporate social responsibility programmes tend to focus on concerns relevant to wealthy individuals and businesses.

These shifts in resource flows have more sharply differentiated the roles of civil society in developing countries. Those engaged in service delivery find it difficult to undertake advocacy that might offend government or philanthropic donors. Those following the social entrepreneurship model are pressed to serve their markets, even if that means ignoring important aspects of social problems they face. Resource streams are associated with specific roles: governments fund service delivery; businesses fund social enterprises; philanthropists fund service delivery or social entrepreneurship. So the shrinking resource base presses civil society organisations to emphasise service delivery and social enterprises. Community mobilisation for self-help, local capacity building, and independent advocacy to improve governance and accountability are victims of the resource crunch.

In industrialised countries the shifts in resources for civil society have taken a different trajectory. Many European countries have long traditions of state funding for civil society provision of many services, as in the Netherlands and the UK (Casey et al. 2010). In recent years, however, neo-liberal approaches have reshaped many social policies. In the UK, the "third way" was seen as a blend between the public and the private financing of social welfare; the current government emphasises a "big society" with less government support (Alcock 2010). Partnership with civil society has gradually shifted to contracting, using market mechanisms to allocate scarce public resources to service delivery (Donnelly-Cox and Cannon 2010). Non-profits that contract for service delivery in UK and Netherlands use market approaches of competition, professionalism, efficiency, and value-for-money. The financial crisis and subsequent recession has further squeezed resources for civil society in many European countries. As unemployment rose, welfare services shrank, and austerity measures emerged, government funding for domestic and international civil society initiatives are increasingly questioned, so many INGOs have fewer resources for overseas programmes.

Shrinking financial resources present significant challenges for civil societies in many countries. Some seek other ways of accessing public funding. Others are realigning themselves to draw greater support from within their societies. Service delivery non-profits have been less affected; funding has been cut for social causes, solidarity issues, advocacy, and promoting diversity (Schulpen 2006). The segment of civil society which works through volunteers and personal donations is less affected. But others may have to choose among publicly funded service delivery, or market-linked social enterprises, or substantial reductions in their budgets and activities. These choices define the functions and roles of civil society more sharply today than in the past decades.

5. Political space for civil society and its relations with political society are simultaneously contracting and expanding

The "Third Wave" of democratisation gave new opportunities for civil society influence two decades ago (Huntington 1992). Citizens began to demand their rights from newly democratic regimes. Many INGOs supported advocacy efforts by civil society actors in developing countries; UN conferences of the 1990s created opportunities for global civil society engagement and alliances. Major international campaigns made civil society increasingly visible as powerful political actors in global arenas and with international institutions (Florini 2000; Fox and Brown 1998).

In recent years, political society (elected representatives, political parties, and public officials) has questioned the legitimacy of civil society roles in public policy and decision making. The Crossroads stories indicate that relations between civil and political societies have become increasingly complex, including dynamics of "constructive engagement", "mutual influence", "mutual castigation and political opposition", and "co-optation". The Chilean student movement, for example, emerged without links to a political party – a break with national tradition. Occupy London took most political parties by surprise, as did the citizen demand for HIV/AIDS treatments in South Africa. In Argentina, support for equal marriage and gender identity laws grew so fast that even movement leaders were astonished by the rapidity of fundamental changes in social attitudes. Many recent initiatives have upstaged political societies, raising issues and mobilising publics to challenge the performance of existing systems.

In some cases civil society actors have collaborated with political society. Some engagements have involved service delivery, as civil society complemented government agencies by organising and mobilising communities to gain services. This collaboration is common in both industrialised and developing countries, though it sometimes threatens civil society independence. Many countries are experimenting with decentralising political authority to local governance institutions, including India, South Africa, Uganda, and Cambodia as well as the UK and the Netherlands. Civil and political societies share an interest in strengthening local governance institutions; many civil society organisations have provided capacity building support to local governance institutions in recent years (Tandon 2011). Civil society contributions to participatory planning and budgeting, for example, have been welcomed and many governments have invited civil society contributions to planning, policy-making, and monitoring activities. Political space for civil society's contributions to government policy and programming has increased in countries such as Chile, Argentina, India, South Africa, and Uganda, as well as in the UK and the Netherlands.

But civil and political society relations can also become contentious: citizen movements challenged government performance in Chile and India; the mineworkers struck against national wage policies in South Africa; farmers resisted government-aided "land grabs" in Cambodia; traders called for citizen action in Uganda against Central Bank policies. It is not surprising that political leaders perceive threats to their position, authority, and legitimacy and respond with hostility. Relations can also sour over issues of accountability. Civil society exposure of corruption is particularly critical of political society, and the reaction is intensified if civil society actors are funded by international sources, as in Russia and India recently. Issues of corruption and lack of accountability mobilise a wide range of citizens, so they can create great insecurity in political society. In response to criticism political society has often maligned, harassed, and intimidated activists. Cambodian security agencies, for example, harass NGOs that provide legal aid to farmers' associations; Putin's government restricts international funding of Russian NGOs that challenge his policies; and ANC leaders in South Africa suppress politically awkward challenges. Using the state security apparatus to intimidate activists has become common in the face of civil society demands for accountability.

Political space for civil society voice and action seems to be expanding and contracting simultaneously. When issues are framed so civil society is seen as an asset to political society, political space expands for citizen action. When civil society challenges public authorities or demands accountability, governments are tempted to shrink the space for that criticism. Citizen distrust of political authority, institutions, and leaders is often linked to emerging activism. The challenge for civil society to frame issues so that political society both appreciates its contributions and tolerates its critiques, enabling civil society to maintain its autonomy at the same time that it influences political society and also holds it accountable.

6. Blurring North-South boundaries reshapes civil society roles and their domestic and international relationships

Historically, domestic NGOs focused on issues of poverty, malnutrition, education, and health care in their own societies. INGOs, mostly based in Europe and North America, used resources from their societies and governments to support civil society work in the developing world (Lindenberg and Bryant 2001). Few INGOs worked in their own countries or explored connections between poverty in domestic and international contexts.

The current blurring of distinctions between North and South poses new challenges for civil societies. Some developing countries have begun to play important roles in the global arena: India, Brazil, and South Africa are active in emerging networks like BRICS as well as global alliances like the G20. Their governments engage in transnational issues like reforming the financial sector, managing fuel and food crises, and dealing with climate change. Private businesses from these countries are increasingly active in transnational commerce. Sometimes governments and businesses stray from domestic norms and policies in their international activities: They exploit resources and labour in pursuit of markets and profits in violation of standards established by decades of struggles within their own civil societies.

Civil societies in emerging countries have been largely unaware of the transnational activities of their governments and businesses. As they become aware, they may be torn between perceived national interests and principles established in global discourses: if their governments reject binding obligations to reduce climate change, should citizens support policies to sustain national economic growth or act to reduce global warming? They may not have access to information about consequences of government and business activities, since public disclosures and debates on such issues are limited in countries like India and South Africa. So globalisation presses these civil societies to build understanding about global issues to complement their domestic experience. There are growing opportunities for sharing civil society innovations and experiences; experience in Chile, Argentina, South Africa, and India is contextually resonant for countries like Cambodia, Uganda, and Jordan.

INGOs that work in developing countries often have limited connections with development issues in their own countries. But the blurring of the North-South distinction is encouraging a new sense of solidarity among civil societies. In North America and Europe citizen movements have challenged the growing income inequality that results from policies that allocate the benefits of economic growth to a few elites. Citizen activists in Europe are learning why structural adjustment and neo-liberal policies generated so much anger and resistance in developing countries. A new sense of solidarity among civil societies in rich and poor countries is emerging as it becomes clear that global elites dominate national and international allocations of wealth and power. Some INGOs are applying ideas from their international work to domestic issues. Dutch INGOs have applied participatory budgeting practices developed in Brazil and India to their local governments, and the UK is deploying community-led monitoring of basic services and participatory urban planning, both pioneered in developing countries. The blurring boundaries also set the stage for wider sharing of innovations and joint learning.

7. Measuring civil society impacts entails expanded definition of long-term success

As public resources for aid and social welfare decline, funders have increasingly demanded immediately visible results and value for money. So success for many civil society organisations has been defined in terms of concrete, often short-term, project goals and results. The indicators and means of measurement are codified in "log frames" approved by international donors, national governments, and private philanthropists. Short-term, project-focused, quantifiable measures of success focus attention on achieving concrete results, but they can also distract attention and resources from larger and more complex systemic changes that are central to the missions of many civil society organisations (Ebrahim and Rangan 2010).

For some Crossroads cases, success definitions emphasised changes for particular constituencies. The struggle over land in Cambodia, for example, sought to protect farmers from exploitation by corrupt officials and buyers. Other cases defined success in terms of changing particular policies and promoting democratic accountability. The traders' strike in Uganda, for example, sought the repeal of a Ministry of Finance policy that made commerce difficult. Progress for particular constituencies or policies often required active citizenship and sophisticated advocacy as prerequisites to success.

In other cases, success was defined more generally in terms of systemic changes in societal priorities and institutions. The campaign to give legal equality to sexual minorities in Argentina defined success in terms of passing legislation, but that legislative change involved altering public attitudes as well as advocacy and legislative campaigning. Some cases focused directly on changing attitudes, norms, and societal institutions. The student movement in Chile campaigned to complete a transition to democratic institutions as well as more equal educational opportunities.

In still other cases, success emphasised citizen expression and innovation for important values not fulfilled by existing institutions. The Occupy and Greek movements gave citizens opportunities to reinvent participatory decision making and mutual education. Russian civil society is becoming a vehicle for citizens' expression and participation in a political system that provides few such opportunities. In measuring the success of these initiatives, the achievement of concrete and measurable goals may be less important than the provision of opportunities for citizen expression, experiment, and participation in shaping their lives.

Why are such expansive meanings of success emerging? At the heart of these initiatives lies the recognition that citizens must engage the growing concentrations of wealth and power within and across societies. Countervailing power is most easily exercised where political space is available, as in the UK or India or Chile. But there are some contexts where political space is shrinking, as demonstrated by the experiences in Greece, Russia, or Uganda (Javeline and Lindemann-Komarova 2010; Sachikonye 2012). Citizen initiatives take on more ambitious goals because of opportunities for enhanced awareness of options, distant solidarities, and coordinated action offered by the revolution in information and communication technology. Fast and interactive communication has enabled citizens to connect with and learn from wide ranges of local, national, and international experiences.

Citizen initiatives for systemic changes that reshape attitudes, norms, and societal institutions can be pivotal for future societies. Compelling historical evidence suggests that pluralistic and inclusive institutions enable widespread participation in economic development and political governance, and that such participation is pivotal to long-term prosperity and accountable governance (Acemoglu and Robinson 2012). Measuring the success of civil society initiatives may require understanding the larger society and its potentials; attempts to quantify short-term outcomes may be useful for some purposes, but assessing large-scale and long-term societal trajectories can be crucial to constructive systemic change. Citizen activism and civil society initiatives that combat widespread tendencies to concentrate power and wealth in the hands of unaccountable elites may be critical to the long-term viability of societies that now struggle with dilemmas posed by economic and political changes.

8. Civil society contributions to institution building invoke complex and conflicted historical processes

The regional viewpoints suggest that civil society initiatives are shaped in part by challenges and opportunities beyond national borders. Citizen initiatives in Europe, for example, responded to the issues raised by financial crises of the European Union and expectations set by welfare states. Civil society organisations long dependent on government funding were ill-prepared to deal with the sudden reductions of government funding and the public energies unleashed by austerity programmes and government responses to unrest. Citizen initiatives struggled to invent new forms and institutions when existing civil society-state arrangements could not cope with emerging governance challenges.

In the Southern Cone of Latin America, citizen initiatives took advantage of the recent return to democratic governance in many countries to launch national debates and legislative initiatives on issues of national rights and responsibilities. The student movement in Chile called for more equity in education; the equal marriage campaign in Argentina and the legal abortion campaign in Uruguay sought to protect rights of disadvantaged groups; the regional "Roof for my Country" movement mobilised youth to house homeless groups. Regional initiatives sought large-scale actions to implement democratic practices and fulfil the promises of emerging democracies.

Citizen initiatives in Asia reflect the diversity of the region. Many Asian countries have experienced dramatic economic growth and associated challenges of resource concentration or corruption. In countries with long histories of democracy and civil society, citizen initiatives have emerged as national accountability campaigns, such as the anti-corruption campaign in India or the railway consumers' movement in Indonesia. In countries with less experience with citizen activism, the citizen uprisings have been more local and episodic, as in the struggles over land grabs in Cambodia, and more outside existing governance institutions.

The countries of the Southern Africa region share a history of colonisation that built extractive political and economic institutions and patterns of violence and impunity from both colonisation and transitions to independence. Civil society organisations in the region have struggled to expand political freedoms and to increase access to wealth, but in many countries new governments have failed to alter existing concentrations of wealth and power. Citizen initiatives have protested abuses of power in countries like Malawi, South Africa, and Uganda – but even short-term successes have been difficult to transform into more inclusive and accountable institutions. Civil society initiatives are increasingly hampered by shortages of support from either local or international resources.

In still other areas, such as the Arab region or Russia, local experience with citizen initiatives has been limited and subject to state control. Initiatives in these regions involved citizen challenges to regimes that were not responsive or accountable to their citizens. Although some of those eruptions of citizen energy displaced existing leaders, they often suffered from lack of citizen experience required to build new governance institutions. If citizen activists cannot construct more inclusive and accountable institutions, new authorities are likely to replicate the centralisation and concentration of power and wealth that sparked citizen protest in the first place (Acemoglu and Robinson 2012).

Building inclusive and accountable institutions is a difficult and demanding process. While some have criticised slow progress in constructing such institutions, evidence suggests that building such institutions has been slow and halting even in countries reputed to be successful (Berman 2012). In regions with longer histories of building such institutions, such as Europe and Latin America and some parts of Asia, citizen initiatives may make rapid progress in promoting more accountable governance. In regions with less historical experience, we can expect more conflict and uneven development as civil societies, markets, and states engage to construct inclusive and accountable institutions that fit regional and national values and contexts, often in the face of determined resistance.

Implications of the crossroads lessons

The lessons that emerge from these cases are relevant to other contexts. The Arab Spring raised similar issues and dynamics; the Occupy movements resonated in many industrialised countries. Thousands of citizen initiatives in response to unaccountable governance are emerging in China. So what are the implications of these initiatives for practitioners and policy-makers?

For practitioners

Civil society practitioners include staff of formal civil society organisations, volunteers in informal associations, activists in a range of campaigns, and many others willing to act on behalf of their values and beliefs to create better local, national and global societies.

Citizen mobilisation, organisation, and conscientisation needs renewed attention: Civil society rests on active citizenship, aware, organised, and vocal. Civil society contributions to citizen awareness and activism have slowed in many countries. NGOs, trade unions, cooperatives, churches, neighbourhood associations, and other civil society groups could play larger roles in enhancing citizen awareness, organisation, and activism.

Formally organised civil society actors need to redefine their identity and mission in relation to citizenship and democracy: NGOs and other organised civil society actors need to distinguish their identities and missions from non-profit service contractors and market-based social enterprises. Clarifying their identities as actors independent of the government and business is critical to articulating the value added by civil society.

Civil society organisations must explore new forms of resource mobilisation: There is an urgent need to expand the resource bases that enable civil society to maintain and pursue its independent mandate. These bases must be linked to the energy, capacity, and creativity of citizens. The lessons from civil societies that transition from social welfare systems may be particularly relevant.

Civil society practitioners need enhanced capacities to operate in ambiguity, uncertainty, and chaos: Civil societies need communications and partnerships with many other actors. They need to use the tools of old and new media to build bridges with dissimilar others, to mobilise energies of youth and other volunteers, and to provide innovative leadership in times of uncertainty, ambiguity, and chaos.

Civil society must promote experimentation and innovation in organising societies: The current model of neo-liberal, market-driven political and economic organisation has created serious problems of inequality, unemployment, and recession. Growth threatens the planetary ecology; democratic governance is vulnerable to concentrated power and wealth, financial instability, and corruption. Civil society organisations can contribute to social learning and innovation needed to solve problems and build better institutions.

For policymakers

The findings also suggest implications for policymakers and donors whose actions can create enabling environments for civil societies to realise their potential for societal contributions.

Recognise the distinctive roles and potential contributions of civil society: Civil society can promote more active citizenship by identifying issues, programmes that do not meet needs, innovations with wider relevance, and unaccountable institutions. Clarifying desired civil society roles and contributions is central to crafting an environment that enables them.

Promote spaces for civil society dialogue with public authorities and private business: Spaces for ongoing dialogues across institutional sectors may reduce "eruptions" from misunderstanding or perceived repression as well as enhance mutual appreciation, understanding, and collaboration on shared agendas. These spaces must include diverse perspectives and operate across many levels to build frames within which collaboration is feasible.

Invest in flexible, responsive, and accountable resource support to civil society: Policymakers and donors need to create flexible and responsive ways to resource civil society, such as incentives for private philanthropy, encouraging citizen participation in voluntary action, or autonomous civil society trust funds. Public resources should support civil society activities that foster public goods and active citizenship.

Support partnerships within and across North-South boundaries for civil society knowledge sharing and solidarity: Policymakers can catalyse large multiplier effects by encouraging more South-South and South-North partnerships that increase knowledge sharing, solidarity, and joint innovation. That support could also build active global citizenship and collaboration to promote shared values.

Invest in long-term capacities for reflection, analysis, and learning in civil society: Opportunities for multi-region reflection on experiences made possible through this Crossroads initiative are rare and precious. Investments can enable civil societies to define and measure impacts, share and sharpen innovations, and promote social learning across practitioners, researchers, and policymakers. In a world where transnational problems and multilevel dynamics pose growing challenges for most societies, investing in learning capacity for sources of social innovation is a high priority.

References

Acemoglu, D., and J. A. Robinson. 2012. *Why Nations Fail: The Origins of Power, Prosperity and Poverty*. New York: Crown.

Alcock, P. 2010. "Building the Big Society: A New Policy Environment for the Third Sector in England." *Voluntary Sector Review* 1 (3).

Badiou, A., et al. 2012. "Save the Greeks from their Saviours!". www.egs.edu/faculty/alain-badiou/articles/save-the-greeks-from-their-saviors.

Beauclerk, J., B. Pratt, and R. Judge. 2011. *Civil Society in Action: Global Case Studies in a Practice-based Framework*. Oxford: INTRAC.

Berman, S. 2012. "The Promise of the Arab Spring: In Political Development, No Gain without Pain." *Foreign Affairs* 92 (1): 64–74.

Casey, J., et al. 2010. "Strengthening Government-Nonprofit Relations: International Experiences with Compacts." *Voluntary Sector Review* 1 (1): 59–76.

CIVICUS – World Alliance for Citizen Participation. 2011. "State of Civil Society 2011." Johannesburg: CIVICUS.

Donnelly-Cox, G., and S. Cannon. 2010. "Responses of Non-Profit Organisations to Altered Conditions of Support: The Shifting Irish Landscape." *Voluntary Sector Review* 1 (3).

Ebrahim, A., and V. K. Rangan. 2010. "The Limits of Non profit Impact: A Contingency Framework for Measuring Social Performance." Boston: Harvard Business School. Working Paper 10-099 from HBS.

Florini, A. 2000. *Third Force: The Rise of Transnational Civil Society*. New York: Carnegie Endowment for International Peace.

Fox, J., and L. D. Brown. 1998. *The Struggle for Accountability: NGOs, Social Movements, and the World Bank*. Cambridge, MA: MIT Press.

Huntington, S. 1992. *The Third Wave: Democratization in the Late Twentieth Century*. Norman: University of Oklahoma Press.

Hupe, P., and L. Meijs. 2000. *Hybrid Governance: The Impact of the Non Profit Sector in the Netherlands*. Rotterdam: Erasmus University Rotterdam.

Javeline, D., and S. Lindemann-Komarova. 2010. "Rethinking Russia: A Balanced Assessment of Russian Civil Society." *Journal of International Affairs* 63 (2): 171–88.

Kilby, P. 2011. *NGOs in India: The Challenges of Women's Empowerment and Accountability*. New York: Routledge.

Knight, B., H. Chigudu, and R. Tandon. 2002. *Reviving Democracy: Citizens at the Heart of Governance*. Rugby: Earthscan.

Lewis, D., and N. Kanji. 2009. *NGOs and Development*. London: Routledge.

Lindenberg, M., and C. Bryant. 2001. *Going Global: Transforming Relief and Development NGOs*. Bloomfield, CT: Kumarian Press.

Ojendal, J., and L. Mona. "Beyond Democracy in Cambodia: Political Reconstruction in a Post-Conflict Society." Thailand.

Papadakis, K. 2006. "Civil Society, Participatory Governance and Decent Work: The Case of South Africa." *ILO Research Series*. Vol. 112, Geneva: ILO.

Sachikonye, L. 2012. *Zimbabwe's Lost Decade: Politics, Development and Society*. Harare: Weaver Press.

Salamon, L. 1995. *Partners in Public Service: Government-Non profit Relations in the Modern Welfare State*. Baltimore, MD: Johns Hopkins University.

Schulpen, L. 2006. "The NGDO Battle – Changes in the Subsidy Structure." *Netherlands Yearbook on International Cooperation*, edited by P. Hoebink. Van Gorcum: Assen.

Sitapati, V. 2011. "What Anna Hazare's Movement and India's New Middle Classes Say about Each Other." *Economic and Political Weekly* 46 (30).

Tandon, R. 2011. "Reforming Governance in India Today." *Yojana* 55: 33–4.

Van Til, J. 2000. *Growing Civil Society: From Non profit Sector to Third Space*. Bloomington: Indiana University Press.

The authors

Rajesh Tandon (corresponding author) is founder-President of PRIA, New Delhi. Over the past three decades, under his leadership, PRIA has extensively engaged in building the capacities of civil society organisations around the world. Rajesh has an electronics degree from IIT (Kanpur), an MBA from IIM (Calcutta), and a PhD in Organisational Behaviour from CWRU (Cleveland). He provided foundational leadership to International Forum of Capacity Building of Southern NGOs (1997–2001), CIVICUS

(1993–1999), FIM-Forum for Democratic Global Governance (1998–2011), and Voluntary Action Network India (1988–2008). He has written extensively, with manuals, research papers, books, and training materials on civil society, capacity development, participatory research, and democratic governance.

L. David Brown is semi-retired as Senior Research Fellow at the Hauser Center for Nonprofit Organizations at Harvard University. He has been President of the Institute for Development Research, a non-profit centre for development research and consultation, and taught at schools of management at Boston University and Case Western Reserve University as well as the Harvard Kennedy School of Government. His research and consulting focuses on civil society organisations, inter-organisational networks, and multi-sector partnerships that foster sustainable development, accountable governance, and social justice around the world.

Index